The Escape from Elba

THE ESCAPE FROM ELBA

The Fall and Flight of Napoleon 1814–1815

NORMAN MACKENZIE

'All my life I have sacrificed everything, tranquillity, interest, happiness, to my destiny.'
  Napoleon to Josephine, March 1807

New York   Toronto
OXFORD UNIVERSITY PRESS
1982

© *Norman MacKenzie 1982*

ISBN 0-19-215863-5

Library of Congress Catalog Card Number: 81-48520

Printed in the United States of America

153260

ACKNOWLEDGEMENTS

I am grateful to Luigi de Pasquali of Portoferraio for advice and assistance, and also to the staff of the public library, the tourist office, and the Napoleonic residences on Elba; to my colleagues Maurice Hutt, who has been characteristically generous with his help, and Giovanni Carsaniga; to the staff of the London Library, the British Library, the Library of the University of Sussex, the Library of the University of Nottingham, the libraries of the Institut Britannique in Paris, the Sorbonne, and the National Maritime Museum; the Historical Manuscripts Commission and the Public Record Office. I also thank Margaret Ralph and Betty ffrench-Beytagh who typed much of the manuscript and Brenda Haw, who drew the maps. The editorial staff at the Oxford University Press are consistently considerate and professionally effective.

The spelling of place names has been standardized to conform with common usage, and silently corrected in quotations. Thus Leghorn is always cited as Livorno, the central square of Portoferraio is referred to as the Piazza d'Armi, even under French occupation, Lyon is spelt as Lyons, Marseille as Marseilles, Genova as Genoa, and so forth. Similarly, frequently recurring personal names, in particular those of members of the Bonaparte family, are Anglicized and unaccented. I would like to express my warm appreciation to the Rockefeller Foundation and to the staff at its Study Center at Bellagio, for the hospitable context in which this book was prepared for press.

CONTENTS

ILLUSTRATIONS

MAPS

CAST OF CHARACTERS

This list is provided for easy reference, and it does not include persons who are of minor importance or those who are given only passing notice in the text.

Captain John Adye	Commanding the British sloop *Partridge* during the escape in 1815
Alexander I	Tsar of Russia from 1801 to 1825
Comte d'Artois	Younger brother of Louis XVIII, who became Charles X in 1824
Giuseppe Balbiani	Sub-Prefect on Elba
Eugene de Beauharnais	Son of the Empress Josephine, a successful soldier and Napoleon's viceroy in Italy
Hortense de Beauharnais	Daughter of the Empress Josephine, wife of Louis Bonaparte, known during the first Restoration as the Duchess of St Leu
Josephine de Beauharnais	Widow of a guillotined aristocrat, Napoleon's first wife and Empress until divorced in 1810
Colonel de la Bédoyère	Commanding the 7th Infantry at Chambéry
Marshal Berthier	Napoleon's chief of staff from 1805 to 1814
General Bertrand	Appointed Grand Marshal of the Palace in 1813, and Napoleon's chief assistant on Elba
Fanny Bertrand	The wife of General Bertrand, and daughter of an Irish general serving in the French army.
Jacques-Claude Beugnot	Royalist Director-General of police
Comte Blacas	Chief minister to Louis XVIII
Caroline Bonaparte	Napoleon's youngest sister, married to Joachim Murat, King of Naples
Elisa Bonaparte	The eldest of Napoleon's sisters, Grand Duchess of Tuscany from 1809 to 1814, married to Felix Bacciochi
Jerome Bonaparte	Napoleon's youngest brother, King of Westphalia from 1807 to 1813

Joseph Bonaparte	Napoleon's elder brother, King of Naples from 1806 to 1808, King of Spain from 1808 to 1813
Letizia Bonaparte	Napoleon's widowed mother, known as Madame Mère
Louis Bonaparte	Napoleon's third brother, King of Holland from 1806 to 1810
Lucien Bonaparte	The only one of Napoleon's brothers and sisters who declined royal honours
Napoleon Bonaparte	Former Emperor of France
Pauline Bonaparte	Napoleon's second sister, separated from her husband Prince Borghese
Louis Guérin de Bruslart	Former royalist resistance leader, appointed military governor of Corsica in 1814
General Cambronne	A veteran of the revolutionary wars, who commanded the detachment of the Imperial Guard which accompanied Napoleon to Elba
Colonel Neil Campbell	British commissioner on Elba in 1814 and 1815
Lord Castlereagh	British Foreign Secretary from 1812, and chief delegate to the Congress of Vienna
General Caulaincourt	French ambassador to Russia from 1807 to 1811, Foreign Minister in the last months of the war, and Napoleon's closest adviser during the abdication crisis
Captain Chautard	Commanding the *Inconstant* on which Napoleon made his escape from Elba
General Dalesme	Commanding the French garrison on Elba at the time of Napoleon's arrival
Marshal Davout	The most loyal of Napoleon's marshals, he continued to hold Hamburg for a month after the Emperor had abdicated
General Drouot	A much respected artilleryman, known as 'the sage of the army', who became governor of Elba and then commanded the Imperial Guard at Waterloo
Cardinal Fesch	Half-brother to Letizia Bonaparte, who owed his successful career to Napoleon's patronage
Fleury de Chaboulon	A young official who took himself to Elba early in 1815 as a self-appointed and clandestine emissary from Bonapartists in France

Joseph Fouché	A notable Jacobin who became Minister of Police and supported Napoleon in his rise to power. A master of secret intrigue, he was again appointed Minister of Police after Napoleon's return to France, but after Waterloo he stage-managed the second Restoration
Francis I	Emperor of Austria from 1792 to 1835, and father of Marie-Louise
Frederick William III	King of Prussia from 1797 to 1840
George, Prince of Wales	Prince Regent from 1811, he was effectively king during his father's long illness. He became George IV in 1820
Baron Hager	Head of the Austrian secret police
Prince Hardenberg	Chancellor of Prussia, and the most trusted adviser of Frederick William III
Hyde de Neuville	French royalist agent, who undertook many secret missions
Comte de Jaucourt	Talleyrand's deputy in the Ministry of Foreign Affairs under Louis XVIII
King of Rome	The son of Napoleon and Marie-Louise, born in 1811, also known as Napoleon II and the Duke of Reichstadt
General Köller	Austrian commissioner accompanying Napoleon to Elba
Colonel Téodor Lacynski	Brother of Maria Walewska, and one of several Polish officers closely attached to Napoleon
Dr Christian Lapi	Commander of the National Guard on Elba
Count Lavalette	A former aide of Napoleon's, married to the niece of Empress Josephine, and Postmaster-General in the last days of the Empire. An ardent Bonapartist
Lord Liverpool	British Prime Minister from 1812
Louis XVIII	Brother of the guillotined Louis XVI, leader of the royalist emigration, he returned to Paris in 1814, and again after Waterloo, remaining king until his death in 1824
Marshal Macdonald	Son of a Jacobite exile, he served through all the wars and played a leading role in the abdication crisis
General Marchand	Commander of the garrison at Grenoble in 1815

Hughes Bernard Maret   One of Napoleon's closest advisers, a trusted diplomat, Foreign Minister from 1811 to 1813, and one of the leaders of the Bonapartist faction in France under the first Restoration

Marie-Louise   Daughter of the Emperor Francis of Austria, she became Empress of France on her marriage to Napoleon in 1810

Chevalier Mariotti   French consul and secret agent at Livorno

Marshal Marmont   One of Napoleon's closest comrades-in-arms, he negotiated the surrender of Paris in March 1814, and planned the defection of the Sixth Corps a few days later.

Comte de Maubreuil   Royalist intriguer, involved in abortive assassination plots against Napoleon

Baron de Méneval   Personal secretary to Marie-Louise

Prince Clemens Metternich   Austrian ambassador in Paris from 1806 to 1809, he became Foreign Minister, and he was the dominant figure at the Congress of Vienna

Marshal Moncey   Commander of the National Guard in Paris in 1814

Marshal Mortier   Commanded the Old Guard during the fighting in Champagne and the defence of Paris

Joachim Murat   Married to Napoleon's sister Caroline, and King of Naples from 1808, he abandoned Napoleon's cause in 1814 after secret negotiations with the Austrians

Count Neipperg   Austrian general and diplomat who negotiated Murat's defection in 1814, and was charged with the supervision of Marie-Louise after Napoleon's defeat

Marshal Ney   Commander of the rearguard during the retreat from Moscow, he was spokesman for the dissident marshals during the abdication crisis and then became a senior officer under the first Restoration

Oil Merchant   The name given to a French secret agent on Elba reporting to Mariotti, who was probably an Italian named Alessandro Forli

Louis Philippe   The duc d'Orléans, who fought with the revolutionary armies in 1793, though a member of the French royal family. He then defected and spent twenty years in exile. Considered a liberal, and an alternative to the more reactionary Bourbons, he was long a focus for intrigue though he was never personally compromised. He became King of France after the 1830 revolution

Guillaume Peyrusse — Napoleon's treasurer

Pons de l'Hérault — Administrator of the iron mines on Elba

Pozzo di Borgo — Corsican exile who became one of Tsar Alexander's advisers

Marshal Schwarzenberg — Commander in chief of the Allied armies from August 1813

General (Count) Shuvalov — Personal envoy of the Tsar and Russian commissioner accompanying Napoleon to Elba

Lieutenant Taillade — Commanding Napoleon's small flotilla on Elba until shortly before the escape

Prince Charles Maurice de Talleyrand — Former bishop, first appointed Foreign Minister in 1797, a strong supporter of Napoleon until 1807. Thereafter much involved in political intrigue, and after secret negotiations with the Allies he became the organizer of the Provisional Government in 1814. A notable champion of French interests at the Congress of Vienna

Traditi — Mayor of Portoferraio

Count Truchsess-Waldbourg — Prussian commissioner accompanying Napoleon to Elba

Captain Thomas Ussher — Commanding the British frigate *Undaunted* in 1814, which carried Napoleon to Elba

Maria Walewska — Known as Napoleon's 'Polish wife', she became his mistress in Warsaw in 1807 and their son Alexander was born in 1810

PROLOGUE

In the first months of 1814 the Emperor Napoleon was defeated, forced to abdicate, and banished to the island of Elba. Within a year, after an escape which startled Europe like a thunderclap, he was back in France and preparing for the campaign which ended in the ruin of all his hopes at Waterloo. There was nothing more remarkable in his whole remarkable career than these rapid changes of fortune, and his response to them, which are the subject of this book.

There have been many biographies of Napoleon, as the legend of his victories echoes down the years, but on Elba we see him at close quarters, stripped of his panoplies and perquisites, yet still displaying the habits of power, deprived of his army and his host of functionaries, yet still moved by the impulses that drove his troops to the Pyramids and to the Kremlin and created an empire that briefly rivalled ancient Rome. Against his two decades of fame the year on Elba seems small, like an image seen through the wrong end of a telescope, but it shines sharp, with every characteristic clear. It shows the egotism of the Corsican soldier of fortune who has tamed a revolution, crowned himself like a Caesar, married a Habsburg, and become the master of Europe. It reveals the brooding patience and the capacity for sudden decision that mark a brilliant commander, the systematic vision of the lawgiver, the benevolence with which he patronizes the arts and sciences, encourages manufacturers, trade, improves agriculture, education, and sanitation, builds roads and harbours. And it focuses on the personal charm and the petulance with which he gets his way, the soldierly camaraderie which seems the only spontaneous aspect of a man so absorbed in his self-created role of greatness that he finds it easier to strike attitudes than to show real pity, anger, remorse or affection.

Napoleon on Elba was indeed a complete miniature of the Man of Destiny, with all his virtues and all his faults scaled down to human size, with no field of action for his abounding energy except the tiny kingdom in which his enemies had with derision installed him; and the contrast between past glories and present adversity was peculiarly humiliating. He had been the foremost man of his time, shaking kingdoms and redrawing maps as the fancy took him, etching such battle-honours as Marengo, Jena, and Austerlitz on the roll of history, turning his enemies into allies and vassals, finding that only Britain stood firm against him, secure from invasion after the destruction of his fleet at Trafalgar, wealthy enough to subsidize a succession of ramshackle coalitions against him. But eventually, by what always seemed to him more a malign stroke of fate than any failure on his part or brilliance on the part of his enemies, the last of these coalitions had beaten him.

The Allied victory, all the same, was a long time coming. Napoleon was the best of generals, commanding what had once been the best of armies, and the retreat which began in the snows of Russia in the winter of 1812 never became a rout as he waged a defensive war all the way back to Paris. As one army was destroyed he found another to take its place. When his opponents outnumbered him he marched his own men twice as fast and made them fight twice as hard. But relentlessly, all through 1813, the Russians advanced across Poland, and into Saxony; the Prussians joined them; and then the hesitant Austrians. At Leipzig in October 1813, in the largest and bloodiest battle of the Napoleonic Wars, the Allies so mauled his army that he was forced to pull back to the Rhine. At the same time, after seven years of dogged fighting in Portugal and Spain, the Duke of Wellington had pushed the French back to the Pyrenees, and a fortnight before the victory at Leipzig he crossed into France.

The battle for France lasted ten weeks, and Wellington said afterwards that Napoleon's campaign in front of Paris was a splendid demonstration of the Emperor's military skills as he feinted and struck, hitting hard at the Russians, turning and moving fast across the battlegrounds of Champagne to check Marshal Blücher's thrusting Prussians, swinging south-east to harass the Austrians, patching gaps in his line, somehow managing to keep control though his opponents were steadily concentrating their forces and he was driven to divide his dwindling armies. But he lacked the strength to win the decisive victory he sought, and the political sense to strike a bargain that would be acceptable both to his exhausted people and to the war-weary Allies. At Prague, at Frankfurt, and most recently in the drawn-out negotiations at Châtillon, he had been offered reasonable terms. Each time he had used the breathing space to regroup for new battles, and at each new instance of bad faith the Allies had raised their price. By the end of February it was clear that they could never conclude a trustworthy peace with the Emperor, and they could do nothing but drive on into France, slowly grinding his armies into regiments of tired and hungry scarecrows, without any notion how the war would end or what would happen in France when it did.

1

THE FALL OF THE COLOSSUS

'If you touch Paris with a finger, the colossus will be overturned', one of Tsar Alexander's advisers said when he advised his master to make a dash for the French capital.

In the last days of a wet and miserable winter the prediction was coming true, for Napoleon had made a disastrous mistake. In the middle of March he had left the road to Paris open while he drove eastwards to take a gambler's chance of cutting off the Allied spearheads. At best there were long odds against him, and the situation was in fact more dangerous than he realized because the Allies had already learned where he was and what he was trying to do. They had intercepted a letter to the Empress Marie-Louise in which he hinted at his plans, and they also had well-placed informants in Paris, including the arch-intriguer Charles Maurice de Talleyrand. Long a key figure in French politics, but long since at odds with the Emperor, he was secretly in touch with the Austrians and he made little effort to conceal his conviction that Napoleon had to be discarded if France was to be saved from a punitive peace. With the Emperor at St Dizier, three days hard ride from his capital, and out of touch with his irresolute brother Joseph who had been left to defend it, there was a chance of defeating him by a sudden blow that was more characteristic of Napoleon himself than of the sovereigns who had come to settle accounts with him. If the Allies were quick they could win the race for Paris.

The Tsar persuaded the King of Prussia and the anxious Schwarzenberg to move forward rapidly, despite the risk of running into a trap which Napoleon could snap shut on them. By 29 March the Allied troops had pushed Marshal Marmont and Marshal Mortier back into Montmartre and the other villages on the northern heights above Paris, defeated the ragged armies of wounded veterans, youths, and civilian volunteers which sought to bar their way into the city, and given the French commanders the choice between an honourable capitulation and—as Marmont put it—'the horrors of a siege' in which the Russians might well take their revenge for the burning of Moscow. Next day they signed an armistice. The French armies were to withdraw from Paris before daybreak on 31 March, and the Allies would make their ceremonial entry a few hours later.

Napoleon raced for Paris as soon as he realized that the main Allied armies had slipped away from him, leaving his own tired units to follow as fast as possible, and hoping that he could reach the city in time to call out the National Guard and to rouse the faubourgs with the revolutionary patriotism that had saved France in 1792. But late in the evening of 30 March, when he reached the Cour de France post-house near Juvisy, expecting nothing except a last change of horses for his carriage before driving on to Paris, he was stunned to see a straggling line of cavalrymen moving south through the darkness and the drizzle. 'What are you doing here?' he asked General Belliard, whose men were leading Marmont's army back to the line of the river Essonne which protected the approach to the royal palace at Fontainebleau. 'Where's the enemy?' he cried in a rage. 'Where are the Empress and my son?' As Belliard reported that Napoleon's brothers Joseph and Jerome had fled with Marie-Louise and the infant King of Rome, and that the capital had been surrendered, Napoleon became distraught. 'Everyone's lost their heads', he shouted at Marshal Berthier, his chief of staff, and General Caulaincourt, his foreign minister, who were travelling with him. 'That's what comes of trusting people who have neither common sense nor energy.' In his distress he started to walk on towards Paris, insisting that something could yet be saved from the wreck of his fortunes, and Berthier and Caulaincourt found it very difficult to persuade him that the capitulation could not be revoked and that he had no means to continue the fight.

For an hour or more Napoleon stumped up and down, declaiming against his brother Joseph and other faint-hearts, blaming himself bitterly for being a few hours too late, devising wild schemes to slaughter the Allies in the streets of Paris, and lamenting that 'the capital of the civilized world will be occupied by barbarians'. Caulaincourt eventually induced him to go into the post-house, but he was so restless with rage and disappointment that nothing would calm him.

He sat for some time in the Cour de France, casting about for a way to turn the fall of Paris to his advantage, for he was certain that the Allies would find it difficult to rally their troops for a battle once they had been dispersed through the streets and squares of the capital. But without an army to hand, and inhibited by the terms on which Marmont had signed the city away, he could think of nothing but the tactics which he had employed all winter. He would send Caulaincourt to negotiate with the Tsar, for the two men had got on well when Caulaincourt was ambassador in St Petersburg, and while he bought time with the talks he would whistle up his last reserves for a gambler's throw. 'Where peace was concerned the Emperor's fairest promise meant nothing', Caulaincourt wrote afterwards, in a telling comment on a man he had otherwise admired and served most faithfully. 'The desire to exalt France's renown and prosperity always came into conflict with his best intentions and most peaceful resolves, and he was forever hoping to escape from the necessity of submitting to a peace that ran counter to his lifelong dreams.'

There never was a clearer example of this trait than Napoleon's reluctance to face the reality of his defeat in 1814. It was nearly dawn when he set off for Fontainebleau, after one of the worst nights of his life, yet his spirits had already risen as he began to improvise schemes to recover his capital. At the age of forty-four, after so many conquests, he could not believe that fate had finally turned against him, and that his refusals to make peace when he had a chance had frittered away an empire and brought the Cossacks clattering down the Champs-Élysées.

*

His wife had already gone the day before he reached the Cour de France, whose very name was a bitter irony in the circumstances. At eight-thirty on the evening of 28 March, when the Russian guns could be heard at the gates of Paris, the Council of Regency met in the Tuileries to decide whether the Empress and the King of Rome should remain in the city or leave it. Although most of the members believed that they should stay, Joseph Bonaparte astonished everyone by reading two letters from his brother: misinterpreted, they were to be the immediate cause of Napoleon's separation from his wife and the loss of his capital.

The pusillanimous Joseph, who had already bolted from his shaky throne in Spain, had long been in favour of making a peace which would save both France and the Bonaparte family fortunes from ruin, and all through the winter he had been upbraiding Napoleon for his reluctance to face reality. Now that disaster was actually at hand he could think of nothing but capitulation and flight, and the two letters provided the excuse.

The first of them had been written on 8 February, when Napoleon was speculating on the possibilities of defeat or death; in such a situation, he told Joseph, his wife and his heir were to be sent to join the troops south of the Loire. 'I should prefer to see my son killed', he wrote, 'than see him brought up as an Austrian prince.' The second letter, couched in much the same terms, was sent from Reims on 16 March. 'You must not allow the Empress and the King of Rome to fall into the hands of the enemy,' Napoleon insisted again. 'If the enemy should advance on Paris in such force that all resistance becomes impossible', he added, the whole government was to decamp to the Loire. 'Do not leave my son, and remember that I would prefer to know that he was in the Seine than held by the enemies of France.'

Napoleon was not dead, the Allied forces which had reached Paris were rashly over-extended, and resistance at a price was still possible, but with such instructions Joseph was able to get his way. In doing so he created so much confusion that he could neither organize a satisfactory defence for Paris nor ensure that the entire government withdrew from the capital. In an attempt to save the Empress and her son, and alarmed about his own prospects, he simply abandoned the Emperor to his fate. 'Well, that's the end of it all,' Talleyrand remarked as the meeting broke up well after midnight, 'though it's throwing the game away with all the trump cards in one's hand.' He had already decided to stay in Paris to see how he could turn his own cards to advantage.

'I'm very upset about going because I know the consequences will be most unfortunate for you', Marie-Louise wrote in a sensible appraisal she sent to Napoleon after the meeting ended, 'but they all said my son would be in danger. . . . So I commit myself to Providence, certain, however, that all will end disastrously.' She had a weak and self-centred personality, and she had been so strictly brought up that she had almost no experience of the world when, for dynastic reasons, her father, the Emperor Francis I of Austria, had married her off to Napoleon when she was only eighteen and her husband-to-be was the most powerful man in the world. Yet she did have a sense of marital and royal duty which made her reluctant to separate from Napoleon at this moment of crisis. She knew very well what she should do, what she probably wanted to do, even though she lacked the strength of character and the means to do it.

Early next morning there was fighting beyond Montmartre as the cavalcade moved off towards Rambouillet on the road to the Loire. There were a dozen *berlines* for Marie-Louise and her attendants, the state coach, and the coronation coach stuffed with household linen;

and behind them rolled and lurched a line of baggage-waggons carrying Napoleon's private fortune as well as the remaining gold of the Treasury, Marie-Louise's jewellery and her wardrobe, Napoleon's robes, uniforms and monogrammed small-clothes. With more than a thousand cavalrymen to provide an escort the hurried departure of the Court was an impressive yet demoralizing sight.

Next day, at Chartres, fifty miles from Paris, Marie-Louise was overtaken by Joseph and Jerome Bonaparte with their wives. The Bonaparte brothers had not even waited for Marmont to settle the terms on which Paris was surrendered to the Allies, and the first definite news they had from the capital was in a note which Napoleon had scribbled at the Cour de France and sent after them by courier. By that time the royal refugees had decided to travel on another seventy miles to Blois. The journey took them two days on roads muddied by incessant rain, and when they reached Blois many wounded soldiers and prisoners had to be turned out to fend for themselves to make room for the Empress, the two former kings (for Jerome had served a term as King of Westphalia at a time when Napoleon was distributing kingdoms to his family), and all their retinues; and the crowding become worse when they were joined by Louis Bonaparte (who had been King of Holland), by Napoleon's mother Letizia Bonaparte, and by her half-brother Cardinal Fesch, who had started life as an army commissary and risen through the Church by his nephew's patronage.

None of them knew what to do. The hypochondriacal Louis, Marie-Louise said later, 'was in such a state of panic that he wanted to leave immediately for some fortress...he was so demented that he was embarrassing', while Joseph and Jerome seemed to think that they would be safer, and have some chance of striking a bargain with the Allies, if they kept control of the Empress, her son, and the Treasury. Napoleon had always been surprisingly indulgent to his brothers, none of whom had merited the honours, wealth, and responsibilities he had heaped on them in his days of glory; and now, in his decline, he was paying the price of leaving them in charge while he was away at the wars.

*

Three of his enemies were no more effective than his brothers. The Prince Regent had played no direct part in the conflict, though the British had been Bonaparte's only consistent foes. The prince's money and marital problems had made him so unpopular that he was hissed in the streets, and even if he had been able to make the journey to France he was physically unsuited to the rigours of a campaign. The Emperor Francis of Austria, who had been twice defeated by Napoleon before he married his daughter to his conqueror, was a peevish, narrow-minded man, obsessed with trivia, fascinated by intrigue, who delegated all important business to Prince Metternich, his able foreign minister, and Marshal Schwarzenberg, the cautious commander-in-chief of the Allied armies. Chased back towards Switzerland by Napoleon's final offensive, Francis was to be the last of the sovereigns to reach Paris. King Frederick William of Prussia was even less impressive. He was more interested in designing new uniforms for his soldiers than in leading them on the battlefield.

The Tsar Alexander, indeed, was the only sovereign with the slightest capacity for leadership in the field, and he had dominated the Allied command in the last weeks of the campaign. An autocrat in Russia, who was to end his days as a mystical reactionary execrated by every liberal-minded person in Europe, his one clear aim was to push his country's boundaries westward as a reward for its sacrifices in the war. At the same time this charming, intelligent and well-educated ruler wished to present himself to the French as a graceful and cultured victor who came merely to restore their liberties. He was seemingly so high-minded that he could dispense with the personal rancour against Napoleon that was so marked among his allies and his own advisers. He had genuinely admired the Emperor in his days of victory, and despite the burning of Moscow and the costly march across Europe he intended to deal generously with him in defeat.

On the morning of 31 March, when Caulaincourt reached the Tsar's headquarters at Bondy, in the outskirts of Paris, he found Alexander undecided what form of government would now be best for France, for the Emperor of Russia had no liking for the Bourbons as a family and no desire to see the obese and politically inept Louis XVIII propped up in the Tuileries by Russian bayonets. He was, however, firm on one point. Napoleon should forfeit his throne. Schwarzenberg was even more emphatic than the Tsar. 'Our objective is to make sure that our children have years of peace and that the world has some repose,' he told Caulaincourt. 'The Emperor Napoleon has shown all too plainly, of late, that he desires neither of these things. With him, there is no security for Europe.'

The armistice had commended Paris 'to the generosity of the Allies', and before their columns marched into the city there were frantic efforts to give France some temporary government. Talleyrand was the moving spirit. He knew that Napoleon was finished: he had been urging him for months to make a peace that would leave France the 'natural' frontiers of the Rhine (including Belgium), the Alps and the Pyrenees. He also realized that there was little hope of a regency for Marie-Louise and her son, for the Allies were bound to see such a solution as an open door for Napoleon's eventual return. Unless France was to become a republic again, or to take some foreign princeling such as the renegade Marshal Bernadotte, who had made himself heir to the Swedish throne, it was clear that the country would have to turn back to the Bourbons. That was an unpalatable conclusion for men who had risen from obscurity in the Revolution, watched Louis XVI and Marie-Antoinette go to the guillotine, and then prospered under the Empire. Yet Talleyrand did not hesitate. Calculating, corrupt, with an unattractive personality and dubious morals, he was nonetheless the most talented politician in France and a consistent champion of the national interest. In this crisis, Caulaincourt said afterwards, 'he gave me the impression of a man compelled by circumstances to marry a girl whom he dislikes and despises'.

When a disconsolate Caulaincourt returned to Paris he thus found Talleyrand trying to form a strange combination of men. There were some who had quarrelled with Napoleon, or lost faith in him; some who were time-servers, hoping to profit by turning their coats; and some were experienced officials who believed that Talleyrand was right, and that the best hope for France lay in ensuring that Louis XVIII returned as a constitutional monarch. It was not a simple task, for the Allied victory had been so sudden that it had caught everyone by surprise, creating such a state of confusion that no one knew what to do next. It was also risky. If Napoleon struck back, and drove the Allies away from the capital, Talleyrand and those who conspired with him might find it hard to keep their heads on their shoulders. There was little public support for the Bourbons, almost unknown after twenty-five years of exile; and the intransigent monarchists, who had no liking at all for a constitution, and despised the notions of liberty, equality, and fraternity, merely wanted to turn the clock back to the days before the people of Paris had stormed the Bastille.

That afternoon the Tsar, the King of Prussia, and Field Marshal

Schwarzenberg led more than 40,000 men into Paris, their bands and banners turning the occasion into a ceremonial parade that was surprisingly well received by the crowds lining the route. The white brassards worn by the troops were solely a means of identification, adopted after Russian and Prussian units had fired on one another by mistake, but they were taken by Parisians to be a show of support for the white flag and cockade of the Bourbon party; and a scattering of royalists took the chance thus afforded to whip up cheers for the king, and then—under the leadership of a hot-headed adventurer known as Maubreuil—to go on to a futile attempt to topple Napoleon's statue from the top of the Austerlitz column in the Place Vendôme. A Russian commander cleared them away and bivouacked his regiment in the square. As the Tsar rode up to occupy his quarters in Talleyrand's fine house in the rue Florentin it was clear that nothing had yet been decided about the throne of France except the removal of its present occupant. 'I haven't seen a friendly face all day', Caulaincourt wrote sadly to Napoleon that night.

★

Forty miles to the south, in a small set of rooms on the first floor of the palace at Fontainebleau, Napoleon was planning to fight rather than submit. As Caulaincourt had already reminded the Tsar there had only been a local armistice in Paris, to which the Emperor was not a party, and Napoleon could easily rally a formidable force along the Loire. He might even make a serious bid to recover Paris. By any conventional standards, of course, Napoleon was not ready for a battle. His troops were weary with weeks of fighting and marching on short commons, much of his artillery, baggage, and other stores had been sent towards Orléans, and for the moment he could count on no more than 70,000 men, less than half the force that the Allies had encamped in and around Paris. Yet the Allied armies were riskily divided. About 50,000 men faced Marmont and Mortier along the line of the river Essonne, which covered the approach to Fontainebleau, 40,000 more were in Paris itself, and another 60,000 were on the right bank of the Seine. It seemed to Napoleon, as he looked at his maps and muster rolls, that a quick blow across the Essonne could send Schwarzenberg's Austrians reeling back into the capital to find that the National Guard and the faubourgs had risen in a patriotic insurrection, while the rest of the French army swung north and east of the city to cut the Allied line of retreat.

In military terms this dramatic move was certainly feasible—so feasible, in fact, that during the first three days of April Schwarzenberg was thinking of pulling out of Paris towards Meaux before the Allied armies were trapped and beaten, one after the other, in Napoleon's usual style; and it appealed to the Emperor so much that he had already drafted the necessary marching orders. But his marshals and divisional commanders did not like the plan at all, for they had no desire to fight a great battle in and around Paris that might turn the capital into a flaming ruin like Moscow whether Napoleon won or lost.

Napoleon had created twenty-five marshals, and though none of them had his political gifts or came near him in strategic brilliance they were brave men in the field, and most of them had served him loyally until his luck turned and he began to throw away French lives and French territory with reckless abandon. Now three of them were dead, some were in disgrace, some were still in the field. Only Berthier, his chief of staff, Ney, Lefebvre, Moncey, Mortier, and Marmont were with the army at Fontainebleau, though Oudinot and Macdonald were expected to come in with the straggling units that Napoleon had left in Champagne when he dashed for Paris at the end of March. It was these eight men who were to play the decisive roles in the next few days, and though they came to look like a group of conspirators they were really no more than a set of tired, disillusioned, and confused soldiers who knew that Napoleon had finally asked too much of his army and that it was on the point of disintegration.

Caulaincourt understood this situation very well. 'The need for rest', he wrote, 'was so universally felt through every class of society, and in the army, that peace at any price had become the ruling passion of the day; and the whole force of this sentiment was levelled at the Emperor Napoleon, whom they accused of having rejected peace, of not honestly desiring it even now.' As such feelings spread Napoleon's power inevitably crumbled. The men who had been driven so long by his restless will, and bewitched by his sense of destiny, were coming at last to the breaking-point, and he could no longer depend on any of them to act for him or to share his magical dreams as they faded in the harsh light of defeat. The most his closest comrades-in-arms could now do for him was to strike the best possible bargain for his life and liberty.

★

For the next two weeks Caulaincourt shuttled between

Fontainebleau and Paris in an attempt to find terms that were accept-able to the Allies and to the Emperor. Everything, in fact, depended on the Tsar. 'Alexander is the most reasonable of them all', Napoleon rightly observed at an early stage in these negotiations. 'Despite all the harm we have done each other he may possibly be the most generous, too.' At the third meeting with Caulaincourt, on 1 April, the Tsar said amiably that Napoleon must abdicate, but he added that the Emperor would be handsomely compensated if he agreed to live outside France or Italy; he mentioned Elba as a possible place of residence, and offered to receive Napoleon as an honoured guest in Russia if Elba did not suit; and he even considered Caulaincourt's repeated suggestion that Napoleon should abdicate in favour of his son. The problem might have been settled in that way if the Tsar and the Emperor had met, as they had met on the raft at Tilsit in 1807, to deal directly with each other.

The Tsar, however, had now to consider his allies, for they had stumbled on to a conclusive victory so unexpectedly that they had arrived in Paris agreed on nothing but the removal of Napoleon from the throne of France—and on the need for strict guarantees that he would never return to it. He was at first sympathetic to the notion of a regency. It would keep out the Bourbons, it would tease the Austrians, with whom he was soon to quarrel about the fate of Poland and Saxony, it would please the French army and most of the people, and since it would lead to a quick settlement it would re-move the threat to Paris. It would also separate the decision about Napoleon's personal future from any peace treaty with France and the arduous business of dividing the spoils of his ruined empire.

But the British were committed to the Bourbons, and they were preparing to send Louis XVIII back from his long exile at Hart-well, a country house north of London; and a regency was the last thing that the Austrians wanted. The Emperor Francis was now as eager to repudiate his son-in-law as he had once been happy to let a Bonaparte marry a Habsburg to found a new imperial dynasty. 'We have only taken this position as a last resort,' Schwarzenberg told Caulaincourt on 1 April; 'now nothing can be changed; we have dealt with you too long, and now we have obligations to our allies. As for the Emperor, he has brought his troubles on himself . . . As for the Empress, her father loves her and we pity her, but the in-terest of Europe and our own repose must come first.' From that moment Caulaincourt understood that nothing could be expected from the Austrians.

Talleyrand and the men he was persuading to form the Provisional Government were equally opposed to a regency. It would have suited Talleyrand very well before the Allies reached Paris, for if he had managed to keep Marie-Louise in the capital he would have been her chief minister and dealt with the Allies, but once she and her son had fled, and he had committed himself to a constitution and a Bourbon king, he could not afford to let Alexander casually ruin his game. He had immediately begun to organize the Senate (where an opposition to Napoleon had been growing for months) to depose the Bonapartes and impose the Bourbons, on his terms. By the evening of 3 April the Senate had done all he wanted. It had passed a long resolution blaming Napoleon for everything that had gone wrong in France, including famine and disease. It had declared that he was no longer Emperor, and that his son had no right of succession. It had firmly declared for a constitutional monarchy. And most important of all, it had opened the way for disaffected commanders to change sides by releasing the soldiers from their oaths.

It was easier to pass such resolutions than to put them into effect. The Allies, or the royalists, might have other ideas. The Comte d'Artois, the reactionary brother of Louis XVIII, would do all he could to save Louis from the constraints of a constitution, and he was already on his way from Nancy with plans that dismayed Talleyrand as much as they would delight the vengeful royalists. Napoleon was also in a position to interfere. Talleyrand had even more reason than the Tsar to fear a swift blow at Paris, and he was certain that neither his government nor any that might succeed it was safe until there was no prospect of Napoleon returning to the capital. There would be no lasting peace, in short, until the Corsican Ogre was banished to the ends of the earth, or dead.

Talleyrand was already speaking of the Azores and St Helena. There was also talk of reviving an old royalist plot to dress a group of reckless men in cavalry uniforms and send them off to find the Emperor and assassinate him. Dalberg, who was one of Talleyrand's colleagues in the new government, was so indiscreet about the scheme that the gossip reached the police in Paris and Napoleon in Fontainebleau. And according to the disreputable Maubreuil, who became involved in this scheme, and then in a protracted series of scandals and lawsuits which suggest that he was mentally unbalanced, it was Talleyrand's own secretary Roux Laborie who suggested that Talleyrand himself had known what was intended.

Whatever the truth of Maubreuil's allegations there is no doubt that Talleyrand was capable of giving covert encouragement to such an intrigue. In the first anxious hours after the capitulation he may well have considered such a plan in case Napoleon attempted to re-take Paris; but with the Tsar ensconced in his own house, and a detachment of Russian imperial guards to protect it, with the Senate decisions in his pocket, and with Napoleon's capacity to make trouble wilting from day to day, the need for such desperate measures soon passed.

★

Unsure what was happening in Paris and unable to exert much influence on events, Napoleon talked endlessly to everyone who came to see him at Fontainebleau, especially to Maret, who had preceded Caulaincourt as foreign minister, and to Caulaincourt himself, who was subjected to a stream of recollections, recriminations and proposals each time he returned to report on his discussions with the Allies. In the course of a few minutes, Caulaincourt said, Napoleon's mood would swing from despair to defiance, from self-commiseration to false hopes that his luck would turn. He said he would abdicate and let the Bourbons sign a humiliating peace dictated by the Cossacks. He declared that he would meet Marie-Louise at Orléans and carry on the war from the Loire, and soon afterwards he was talking of joining his stepson Eugene Beauharnais in Lombardy and making a last stand as King of Italy. He complained continually about Talleyrand's career of treachery and about the gross ingratitude of the crowned heads of Europe, whom he had snatched from the abyss of revolution. He continued to insist that when the Emperor Francis eventually reached Paris he would put everything right for the sake of his daughter.

Napoleon had always been a man of impulses. In this state of shock, however, his natural restlessness was so accentuated that one impulse followed another without leading to any decision. In all these flights of ideas, in fact, the only consistent theme was his desire to fight an apocalyptic battle for Paris, and though that prospect alarmed the men who would have to plan and fight such a desperate venture Napoleon continued to delude himself by appealing to the most devoted of his soldiers. On the morning of 3 April, when he reviewed the Guard in the courtyard at Fontainebleau, he was so gratified by the shouts of 'To Paris! To Paris!', and the fervid singing of the 'Marseillaise', that he spent the afternoon with

Berthier drafting a plan to retake the capital two days later.

During the night, however, the broken units led by Macdonald and Oudinot began to reach the shelter of the Forest of Fontainebleau. It was clear to everyone that the young conscripts were in no state to fight, and unlike the veterans of the Guard they wanted peace. General Gerard, who commanded the Second Corps, told Macdonald bluntly that his army was falling to pieces, and that the generals would refuse to march on Paris and add the destruction of the capital to so many other sacrifices.

Next morning, after another review, Berthier, Maret, and Caulaincourt were talking to Napoleon when they were joined by Ney, Moncey, who commanded the National Guard, and Lefebvre, who had plucked up their courage to tell Napoleon that the situation was hopeless. Napoleon began to chide and ridicule them, as he had done so often in the past when they differed with him. 'We will save France, redeem our honour, and then I will accept a moderate peace,' he declared in a flurry of familiar rhetoric. 'A last effort, and then we can all relax after twenty-five years of struggle.'

At this point Macdonald arrived with a depressing report on the state of his troops, saying that they shared the widespread fear that Napoleon would expose Paris to the dreadful fate of Moscow, and showing Napoleon a letter inviting him to defect to the Provisional Government. As the argument went on it became clear that the marshals would only be satisfied if Napoleon agreed to abdicate in favour of his son. If the fighting started again they would face defeat and disgrace, or victory and endless war; and if the Bourbons returned they could expect little from the dynasty they had been fighting since the Revolution. But a regency under Marie-Louise would seemingly give them exactly what they wanted—peace, and a continuing share in the power and perquisites of the Empire.

Napoleon thought them fools as well as faint-hearts. 'The marshals have lost their heads', he said to Caulaincourt after he had failed to convince them that a regency would not last a week. 'They can't see without me there will be no army, nothing to guarantee their interests.' All the same, he was obliged to make a show of compliance. 'Very well. If that's how things stand I abdicate,' he said with petulant self-pity. 'I have sought nothing but the happiness of France, and I have failed.'

Even then he was temporizing, for he was a man who liked to insure against all contingencies. He sent the marshals away while he drafted a conditional abdication, but as he talked to Caulaincourt about the ensuing negotiations with the Tsar he kept reverting to the idea of seeking his fate under the walls of Paris, and insisting that his reluctant marshals would have to fight once they were committed to a battle. The point was so strong in his mind that he could not conceal it when he recalled them. After he had read them the draft abdication and told them that he was sending Caulaincourt, Ney, and Marmont to discuss it with Alexander, he spoiled their relief with a teasing appeal to them as old comrades-in-arms: 'Let's march tomorrow and beat them,' he cried. 'Be ready to move at four in the morning.' The marshals were silent, and after a few moments Napoleon irritably scratched his name on the paper.

Such ambivalence was characteristic of Napoleon's behaviour all through the first week of April, and it is impossible to say whether he wanted the negotiations to succeed or whether he hoped that the Allies would reject the idea of a regency and thus force the marshals to stand by him. He may not even have known himself, for he said so many different things in the course of this confusing day that the commissioners were glad to get away with the note of abdication in their hands—as well as another vague document saying that once the principle of a regency had been accepted they could go on to sign a treaty of peace.

This uncertainty, moreover, was reflected in another apparently trivial but actually crucial change of mind. Before the commissioners left Macdonald was told to take Marmont's place with Caulaincourt and Ney. This was understandable, for Napoleon had known and trusted Marmont since they were young officers together and it was the Sixth Corps, which Marmont commanded, which would have to bear the brunt of an Austrian attack or lead Napoleon's march on Paris. Yet at the last moment, possibly prompted by fears about Ney who was doubtfully loyal and such a fool in politics that Talleyrand might easily play on his vanity and ambition, Napoleon told the three commissioners that they could decide when they got to Essonnes if it was sensible to take Marmont with them.

It was a very curious decision, for Napoleon had already decided to call a meeting of staff officers next morning to consider plans for a battle if the commissioners were rebuffed, and for that purpose Marmont was the most necessary of them all. It was also a disastrous decision. While Napoleon hedged and hesitated at Fontainebleau his old friend had been blundering into an intrigue that would destroy all chance of a regency, or a victory over the Allies.

★

Talleyrand was well aware that the marshals wanted peace, but he could not afford to let them settle for a regency. He therefore began a systematic attempt to subvert them, sending old friends-in-arms through the lines with letters urging them to avoid the horrors of a civil war and telling them that the Senate had released them from their oaths of loyalty to the Emperor.

The Senate resolution was the trump card. It was to have a rapid and general effect on the French army, for it allowed anyone to change sides without too much fear of the consequences; and if there was going to be a new government the sooner one took service under it the better. It also had a particular effect on the marshals, who had most to lose or gain in the matter. If they had been threatened with attainder, or simply told that there would be no place for them in the new order of things, they would have been obliged to take their chance with Napoleon to the end. But Talleyrand's ingenious resolution had assured them that it was safe to defect, and that if they put their loyalty to France before their loyalty to the person of the Emperor they might hope to continue their careers, to retain their titles, their social position, their fine houses and country estates even under a Bourbon restoration. It was tempting bait for simple and self-important men who had spent much of their lives campaigning, and it was offered at exactly the right moment, when the marshals saw that Napoleon had finally led them into an impasse.

The seductive messages were therefore sent to almost all the marshals, but Marmont was obviously the prime target. If the Sixth Corps could somehow be spirited away from Essonnes the Provisional Government would be safe and the Allies could impose what terms they wished on Napoleon.

In the late afternoon of 3 April a former aide of Marmont named Charles de Montessuy arrived at his headquarters dressed as a Cossack and carrying a flag of truce. He had brought the Senate resolution, letters from four close acquaintances who had joined Talleyrand's government, and an appeal from Schwarzenberg; and he added his own persuasions. Marmont wavered for an hour or so, but he was in no mood to resist these blandishments. He was exhausted by weeks of continual fighting, he was uncertain what was happening at Fontainebleau, he was afraid of a military collapse degenerating into civil war, and he feared that he had already compromised himself with Napoleon by surrendering Paris. It is also possible that he saw himself as the saviour of France, and the means of protecting the Emperor from ignominy or death. Whatever his motives he made no attempt to inform Napoleon of these approaches, and it was entirely on his own initiative that he told Schwarzenberg that he would capitulate on two conditions. His army must be allowed to retreat to Normandy with all its supplies, and the honours of war; and if, as a consequence, Napoleon was to be captured by the Allies they were to guarantee his life and his liberty in a place they would choose in consultation with the new government of France.

Early on 4 April, while Napoleon was reviewing his troops at Fontainebleau in complete ignorance of these exchanges, Schwarzenberg sent a cordial reply saying that he accepted the proposed terms and 'appreciated the delicacy' of the clause about the Emperor. At the same time he told his generals to be on their guard as Marmont's corps marched through the Austrian ranks that night on their way to Versailles. Even if Marmont was in earnest it was possible that his troops might break loose when they found themselves betrayed, or that Napoleon might use such a complicated manoeuvre as cover for a surprise attack.

It was four in the afternoon when Caulaincourt, Macdonald, and Ney reached Marmont, and their unexpected arrival understandably made him agitated and uncomfortable. After some hesitation he told them that he had been corresponding with Schwarzenberg, but he claimed that he was still waiting for a formal agreement. If nothing was yet signed, Caulaincourt said firmly, nothing was yet settled; and so, after some further discussion, Marmont agreed to go with the commissioners to tell Schwarzenberg that the Emperor's conditional abdication had made the proposed surrender unnecessary. 'I no longer had any need to incriminate myself,' he said afterwards. 'My duty commanded me to join my comrades. I should have been guilty if I had continued to act alone.' Though he had already told his divisional commanders that he intended to capitulate that night, and found all but one in agreement, he now reversed himself and before he left he gave his deputy General Souham strict instructions that no one was to move until he returned or sent fresh orders.

When the commissioners arrived at Schwarzenberg's headquarters they were given a chilly reception and the Allied commander-in-chief showed no interest in their idea of a regency. 'We must make peace when it's possible', he told Caulaincourt. The much-embarrassed Marmont had a private interview. 'I released myself from the arrangements that we had made,' he claimed later. 'He

understood me perfectly, and gave complete approval to my plan.'
Yet Schwarzenberg seems to have said nothing about Marmont's
proposed surrender to the other commissioners before he sent them
on through the night to Paris.

The members of the Provisional Government, of course, had
been profoundly relieved when they heard that Montessuy had
talked Marmont into surrender, and they waited anxiously through
the evening of 4 April for the news that the units of the Sixth
Corps were actually crossing into the Austrian lines. 'Here is a
piece of good news, my dear,' Talleyrand wrote with relieved satisfaction to the Duchess of Courland sometime that day. 'Marshal
Marmont and his corps have just surrendered, an event which is
due to our proclamations and writings.' The relief was premature.
Talleyrand was stupefied when a courier arrived to say that Napoleon's commissioners were on their way with an offer to negotiate.
This was the last thing that Talleyrand wanted, yet he could not
prevent the Tsar from receiving the delegation quite amiably when
they arrived at the rue Florentin in the small hours.

Far from surrendering, Marmont had turned up in Paris, and he
clearly gave his colleagues his full support. 'I was not the least ardent in defending the rights of the son of Napoleon and the Regent', he said. But once again Alexander rejected the proposal to
set up a regency. 'It's too late', he told the commissioners, 'and
things have moved too fast.' All the same, he repeated that he had
no desire to impose the Bourbons or any other form of government
on France, and Caulaincourt had the impression—as the commissioners left at three in the morning to the accompaniment of
acrimonious exchanges with the waiting members of Talleyrand's
clique—that the Tsar might still be willing to accept a regency.
After Napoleon's emissaries left, indeed, Alexander made a final
attempt to persuade Talleyrand. He would not give way. 'If you set
up a regency', he told the Tsar emphatically, 'then Napoleon will
be back in a year.'

Away to the south, moreover, events had been running faster
than the commissioners could travel and talk. When Napoleon's invitation to Marmont for the staff meeting next morning reached
Essonnes and was followed within the hour by Colonel Gourgaud,
sent to make sure that Marmont and his generals would attend, it
seemed to the apprehensive General Souham and most of his colleagues that the invitation was nothing more than a trick to lure
them all to Fontainebleau. Fearing that Napoleon had somehow

learned of their proposed defection and that he might be planning to put them in front of a firing squad, they at once decided to cross into the Austrian camp. Marmont, they decided, had already bolted to save his skin. A few of them refused to panic and left for Fontainebleau, while Colonel Fabvier rode hard for Paris to warn Marmont. But it was too late. By dawn Marmont's troops were surrounded by their enemies.

The commissioners had slept late, and they were still at breakfast in Ney's house when Fabvier came in with the disastrous news. Marmont seemed genuinely surprised and distressed that his anxious subordinates had actually carried out his plan to defect. 'I would have given an arm to prevent this,' he said glumly. 'Even your head wouldn't be too much', Ney replied, guessing at Napoleon's reaction and the reception they would receive when they went back to see Alexander.

The defection of the Sixth Corps had indeed transformed the situation and changed the mind of the Tsar. There was no longer any need for him to strike a quick bargain with Napoleon, or to object to the return of the Bourbons if Louis was constrained by a constitution. 'The Emperor must now abdicate unconditionally', he told Caulaincourt that morning—though he promised that Napoleon should be given some still unspecified kingdom of his own. Corsica had to be ruled out because it was French, Sardinia because it was part of Savoy, Corfu because the English and Russians would object; and even Elba, which Alexander himself favoured, was rather too close to Italy for comfort. All that the Tsar would offer, in the immediate aftermath of the devastating news from Essonnes, was a two-day armistice and a pledge to discuss Napoleon's personal future sympathetically if Caulaincourt returned with the Emperor's unqualified abdication.

Napoleon had prevaricated too long, as even his closest adviser realized. 'Fearing to fall into the hands of the Allies', Caulaincourt wrote afterwards, 'he played his game with so much anxiety and indecision that he lost his supporters, gave opportunities to his French enemies, and in the end had to depend for his rights and security on the men who had beaten him.'

2

SURROUNDED BY WOLVES

About two o'clock in the morning of 5 April, when Captain Magnien woke Napoleon with the news that Marmont's corps had gone over to the enemy, the Emperor immediately sent a cavalry unit to cover the gap until Mortier could move across to fill it. There could now be no question of attacking Paris. It was even doubtful whether Napoleon could hold his position at Fontainebleau for more than a day or two if the fighting started again, for the Allied units were still pushing forward and threatening to cut him off from the south and the west. Yet he still felt that a longer war was better than a shameful peace, and when Ney returned from Paris he found that once again the Emperor was talking of falling back on the Loire, and that Berthier had already been told to prepare orders for a march in that direction.

Ney had hurried back to Fontainebleau, apparently hoping that he might get the credit for persuading the Emperor to abdicate without any conditions; he had been the most outspoken of the disaffected marshals, and he now wanted to trump Marmont's spectacular defection by taking Napoleon's signed abdication back to Paris. He told Napoleon bluntly that there was now no hope at all of further resistance, and he was so convinced that he had persuaded him to give up the throne that at 11.30 p.m. he wrote a jubilant letter to Talleyrand. 'The Emperor is resigned to his fate', he declared, 'and is ready to agree to a complete and unconditional surrender.'

Ney was mistaken. When Caulaincourt and Macdonald arrived just after midnight, bringing the Tsar's demand for abdication and his offer of Elba, Napoleon greeted them with a series of tirades—against Talleyrand, who had been planning for years to deceive him, against Marmont, who had deserted 'when victory was at hand', against the cowardice of his brother Joseph, who had lost him Paris, and against all those who were abandoning him for 'interest, place, money and ambition'. He seemed to shrug off any responsibility for his downfall, and to see himself as the victim of fools and malign adversaries.

It was almost dawn before Caulaincourt could calm him and get him to draft a new and unconditional abdication, which was to be handed over on signature of an acceptable treaty. 'Since the Allied Powers have proclaimed that the Emperor Napoleon is the only obstacle to the restoration of peace in Europe', the document read after some small alterations to the first scribbled phrases, 'the Emperor Napoleon, faithful to his oath, declares that he renounces for himself and his heirs the thrones of France and Italy, because there is no personal sacrifice, even of life itself, which he is not prepared to make in the interest of France.'

Only three hours later, however, Napoleon was having second thoughts, and he called in Ney, Macdonald, and Caulaincourt for yet another discussion of a march to the Loire, and the even rasher alternative of a dash to join Eugene in Italy. Once again they pointed out the risk of a military disaster degenerating into a guerrilla war, and once again he upbraided them. 'Very well. If you want rest you shall have it. But you don't know what dangers and disappointments you'll find when you lie on your feather-beds. A few years of such a peace, for which you'll pay dearly, will make you wish you'd fought a most desperate war instead.'

This round of ranting persuaded Ney and Macdonald that Napoleon was now so desperate that he might order what remained of his army into some such forlorn hope and wreck this last chance of a reasonable peace. When they left him, therefore, they talked to the other senior officers still at Fontainebleau, and found them all ready to strip the Emperor of his remaining powers. The commissioners who were acting on his behalf would now make all the decisions for him, and in the meantime they told Berthier to ignore any military orders except those they might issue to support their negotiations with the Allies.

Napoleon had thus become an Emperor who had lost his empire, a great general who was reduced to commanding a palace guard, a politician without a policy, and a virtual prisoner of his own senior officers. He could not even expect to bargain much, either for France or himself, though he had a lingering and groundless hope that once he abdicated the Allies would let France retain Belgium and the Rhineland, and he kept repeating that if the country were deprived of these natural frontiers there would be a new war within the year. But before Caulaincourt went back to Paris the Emperor told him that he must not release the letter of abdication until he was offered satisfactory terms, including a firm offer of a personal

kingdom on Elba, adequate pensions for his family and other de-
pendents, Tuscany for Marie-Louise, and permission for the Empress
and the King of Rome to join him at Fontainebleau before his departure. He was whistling in the dark. In settling their accounts
with him the Allies were also proposing to wind up his dynasty for
good.

★

Despite the Tsar's promise to safeguard Napoleon's personal interests there were continuing difficulties about Elba and Tuscany,
which had reverted to Austrian occupation. 'The question of the island of Elba arouses discussions,' Talleyrand wrote on 7 April.
'The moral condition of Italy does not seem to admit of such an
establishment'; and before the treaty was signed he threatened that
unless Napoleon was sent right away from Europe the Provisional
Government would not pay him any pension or recognize him in
any way. 'It is deeply to be regretted that Lord Castlereagh was not at
hand', his half-brother Sir Charles Stewart wrote to Lord Bathurst
in London, 'to counterbalance by his moral resolution and strong
sagacity the imprudent and somewhat theatrical generosity of the Emperor Alexander.' It would be well to consider, Stewart added,
'whether a far less dangerous retreat might be found, and whether
Napoleon may not bring the powder to the iron mines of Elba'.

Castlereagh in fact reached Paris on 10 April, when the Treaty of
Fontainebleau was drafted but not yet signed. He at once declared
that he disapproved of it and refused to sign it on behalf of his government, but he reluctantly waived his objections to the choice of
Elba because the Tsar's word had been pledged; 'the sovereigns',
he said a year later when his critics asked why he had agreed to
such muddled and unsatisfactory arrangements, 'thought it the only
way to bring the conflict to an end'. The Prime Minister, Lord
Liverpool, offered the same apology a year later when he too was
asked why the Allies had failed to make 'due provision' against
Napoleon's return to France. The Treaty of Fontainebleau, he insisted, 'afforded the only means of avoiding a civil war in France,
and of bringing the marshals over'. If the war had gone on, he
added, conceding Napoleon's point that the treaty had been a bargain and not a punitive sentence, the Allies might have been successful, 'yet the struggle would have been formidable, and its result
doubtful'.

At the same time that Castlereagh was complaining that the

proposed treaty was too loosely drawn and indulgent to the Emperor, and Metternich had arrived in Paris to press the Austrian objection to his installation as King of Elba, Napoleon himself apeared to be changing his mind again. His senior officers were drifting away all the time, pleading family and other business, but his spirits were still buoyed up by the cheers of the veterans he called his *grognards*, or old grumblers, and by his illusions. He now threatened to withdraw his offer of abdication and to revive his plan to link Augereau's troops with the army of Italy, and he justified this ludicrous notion by a piece of gossip about his father-in-law which seemed to suggest that Francis might after all decide to save the throne of France for his grandson. Caulaincourt was horrified. 'We are already regarded as rebels', he wrote, begging Napoleon not to make double-dealers out of the commissioners as they struggled to do their best for him. 'There is nothing to be hoped from Austria', he added, explaining that even with the Tsar's support he could not obtain Tuscany for the Empress. This reply did not calm the Emperor, who sat down at five o'clock in the morning of 11 April to write Caulaincourt an angry letter about the Austrian treatment of Marie-Louise, and told him bluntly that he was not to sign the treaty unless there was an agreement on Tuscany.

Napoleon now put forward a completely different idea. He told Caulaincourt to go to Castlereagh to ask if he could be given asylum in England, without the need for a treaty, and in complete liberty. After twenty years of war Castlereagh was naturally astonished to receive such a request. Napoleon was in effect suggesting that he should change places with the returning Louis XVIII, making it clear that he trusted the word of his most obdurate enemies more than the promises of the men with whom he was currently negotiating his fate. But it was too late to change anything. By the time Caulaincourt received this letter he and the other commissioners had already signed the document and if Napoleon had repudiated it he would have become an outlaw.

The most important of the twenty-one articles in the Treaty of Fontainebleau dealt with titles, pensions, and possessions. After noting the Emperor's unconditional abdication it promised that he and his family should retain their titles, that he should have 'full sovereignty and property' over the island of Elba, together with a pension of two million francs a year from the French government; that the Duchies of Parma, Placentia, and Guastalla (across the Apennines from Tuscany) should be settled on Marie-Louise and her son; that an annual pension of two and a half million francs should be divided between Napoleon's mother, brothers, and sisters; that the interest on a further two millions should be paid to such of Napoleon's generals and employees as he chose; and that generous provision should also be made for the divorced Empress Josephine and her son Eugene. If the French government were to carry out all the obligations being laid upon it by the Allies—to which Talleyrand was obliged to give his formal consent—the support of the fallen Bonaparte family would cost it the equivalent of £300,000 a year.

With the exception of Tuscany, Napoleon had thus got more or less what he had asked from the Allies, and most of the remaining articles dealt with the protocol and practical matters involved in setting him up with all the dignities of a petty prince. He was offered support against the depredations of the Barbary pirates, who had long raided Elba and seized Elban ships. He was to be given an escort to the Mediterranean, ships to carry him and his household over to his new capital at Portoferraio, a naval corvette to serve him afterwards, and 400 men picked from his Imperial Guard to defend his little kingdom.

The treaty as a whole required very little from Napoleon in exchange for all these concessions. There were the arrangements he had himself proposed for transferring his personal properties and the state jewels back to the Crown, and for the return of the Treasury gold which had been carried off in Marie-Louise's baggage-train. But there were no clauses requiring him to be tranquil on his island or stipulating penalties for new intrigues or adventures: he was not even told that he must remain on Elba or forbidden ever to return to France. The 'strict conditions' which Alexander had mentioned to Caulaincourt at their first meeting seemed to have been forgotten in the rush to get Napoleon away from Fontainebleau before Louis XVIII reached Paris and the Allies turned to the task of making a more general peace treaty with France.

The situation, indeed, was becoming embarrassing. On 10 April there had already been a ceremony of 'purification' on the place where Louis XVI and Marie Antoinette had been guillotined. And the timing was so tight that on the morning of 12 April, as Caulaincourt and Macdonald rode out of Paris to the south, their colleague Marshal Ney was hastening out of the north side of the city to greet the Comte d'Artois at Bondy. The most dashing of Napoleon's marshals had gone straight from signing the treaty in his name to a meeting with the man who was brother both to the dead king and to the new one.

★

The position of Marie-Louise was now most peculiar. It might be assumed from the treaty that Napoleon would be able to take his wife and child with him into exile, or that they would be free to visit him in Elba. But there was nothing in the text to justify that view. The only relevant clause spoke of issuing passports to the Emperor, the Empress, and other members of the family 'who may wish to accompany them or to establish themselves outside France'. The Austrians had cleverly managed to avoid any commitment to bring the Emperor and Empress together again.

In the first few days after the collapse Marie-Louise certainly thought of setting off to join Napoleon at Fontainebleau, and she could probably have made the journey of over a hundred miles quite safely. 'I had the honour of seeing her very often at this painful moment', Baron de Méneval, her loyal and conscientious secretary said, 'and I became convinced of her devotion to the Emperor.' But everyone was playing on her fears with tales of Cossacks raping and looting across the countryside, and Napoleon himself made no effort to persuade her to leave Blois. At first he clearly thought it better to leave her where she was, with his family and the state treasure, in case he suddenly decided to bolt for the Loire. Then, on 7 April, a Colonel Galbois reached Blois with a letter in which Napoleon told his wife that the Allies were talking of sending him to Elba, complained that the Austrians were objecting to the cession of Tuscany, and said that the Tsar had sent a Russian officer to escort her to Fontainebleau. 'I'm sad', he concluded, 'that all I can do for you is to oblige you to share my ill-fortune.'

This was the first Marie-Louise had heard of the abdication or of the proposal to exile Napoleon from the mainland of Europe, and her immediate reaction was sympathetic. 'My place is at the Emperor's side when he is unhappy', she told Galbois, but when he added his voice to those which urged the perils of travelling on roads filled with marauding troops she agreed to wait until the promised Russian officer arrived.

The prospect of her departure, however, so alarmed Joseph and Jerome that it brought them to a decision, and early next morning, on Good Friday, they entered Marie-Louise's apartments dressed in travelling clothes, saying brusquely that the Russian army was close to the town and that they must all leave at once. According to one account they proposed to set up a new seat of government some-where south of the Loire, but according to the letter Marie-Louise sent to Napoleon, protesting against the behaviour of his brothers, they had insisted 'that I must hurry away to hand myself over to the nearest Austrian corps'.

Whatever the precise proposal may have been it provoked a bit-ter argument marked by the threats and shouting that characterized rows in the Bonaparte family, and Marie-Louise told Napoleon that the officers of her guard made it clear that 'they would not tolerate my being obliged to leave . . . and that if your brothers were fright-ened all they had to do was to get out as soon as possible'. In this shaming scene the Empress seemed the one calm and reasonable member of the imperial family. 'These tantrums cannot affect me,' she wrote to Napoleon. 'I am waiting for your orders and implore you to send them to me.' All the same, behind her pose of dignity she was sufficiently upset by the bullying of the Bonaparte brothers to write to her father asking if she could find asylum in Austria. 'All I hope is to live quietly in your dominions, no matter where', she wrote, 'so that I may bring up my son.'

At two that afternoon, however, the expected Russian officer ar-rived, introducing himself as Count Shuvalov. There is no doubt that the Tsar had instructed him to take Marie-Louise to Napoleon, for she wrote that day to her father saying that Shuvalov had ex-plained 'the situation in which the Emperor finds himself' and that she would 'leave for Fontainebleau tomorrow'. It is equally clear that Francis already had other plans for her. As soon as he heard about Shuvalov's mission he protested that the Allies had taken a decision 'with regard to the person of his daughter' without con-sulting him, and he demanded that she 'should be sent to him, so that he may conduct her, in a manner worthy of her birth, to his dominions, and that he may give her and her son a suitable estab-lishment until the time comes for her future to be decided'.

Francis was a high-minded hypocrite to whom duplicity came easily. 'I've no complaints to make about my son-in-law', he wrote to Marie-Louise, insisting that he wished to see her marital happi-ness continue. But in this letter he carefully distinguished between his fond feelings as a father and his formal commitments as an emperor, obliged to accept the decisions of his fellow sovereigns; and at the same time he wrote to Metternich making it clear what he wanted those decisions to be. 'The important thing is to see that

Napoleon is banished from France, and please God, the further the better. I don't approve of the choice of Elba as Napoleon's residence. It's a part of Tuscany, and this means that some of my territory is being disposed of in favour of a foreigner. I really cannot permit this sort of thing in future.'

Shuvalov's arrival was enough to persuade most of the functionaries who had followed Marie-Louise to Blois that the Empire had finally collapsed, and as soon as the Russian had given them safe-conducts and they had paid themselves out of the public funds they left to try their luck with the new government in Paris. Shuvalov, meanwhile, managed to get the Bonapartes on the move again and that night—after Shuvalov had been forced to drive off an attempt by Cossacks to plunder the baggage and treasure waggons—the whole party reached Orléans.

Soon after their arrival the situation was complicated by a letter from Napoleon to Méneval in which he said that the Austrians had agreed to the succession of the King of Rome, with Marie-Louise as regent, and that she should therefore look to her father to protect her interests. The first part of this letter was an illusion; and the second was unfortunate, for it again told Marie-Louise to count on her father rather than her husband.

She was understandably confused as messengers were delayed and letters crossed, in a situation which was changing from one day to the next. Perhaps the Bonaparte brothers had frightened her, or she mistrusted Shuvalov. Whatever the reason she now sent one of her officers with a letter to Francis asking for permission to join him. 'The Emperor is leaving for Elba', she wrote, 'and I have told him that I cannot go with him until I have seen you and heard what you have to say. ... Every day my position becomes more critical and more alarming. There is a plot to force me to leave against my will. I implore you to send me your answer as soon as possible, because I'm nearly dead with fear.'

Both the trappings and the substance of the Empire were fast slipping away from her. The Provisional Government, desperately short of cash and afraid that the equally impoverished Bourbons would get their hands on the money first, had not waited for the Treaty of Fontainebleau to provide for an orderly return of the Treasure, or for Napoleon to keep his substantial private savings. It had sent a disreputable official named Dudon to lay his hands on whatever funds he could find in the imperial baggage-train. When Dudon reached Orléans on Easter Sunday he not only took the stocks of gold that properly belonged to the state, he also seized a large part of Napoleon's own fortune, his snuff boxes, uniforms and personal linen, and much of Marie-Louise's own jewellery— including the string of pearls that she was wearing at the time. All that was left was a sum of about six million francs which was treated as her personal property.

Next day there was another letter from Napoleon, asking Dr Corvisart (the able court physician who had gone with Marie-Louise to Blois) whether she would be well enough to go to Italy and take the waters at some spa, such as Lucca or Pisa, which would be convenient to Elba. Corvisart, however, had no desire to go to Tuscany, let alone share the fallen Emperor's exile, and he may have been acting as a paid agent for the Tsar or for Schwarzenberg: possibly both. It was certainly easy, with such a patient as the Empress, for him to find sufficient symptoms to block Napoleon's proposal. Marie-Louise, he declared, was subject to pulmonary weakness, headaches, and nervous debility and since she was 'habitually poorly' in the middle of the month there could be no question of her travelling southwards with the Emperor as he made his way to Elba. What she needed, he said, was 'absolute rest and tranquillity in some suitable spot where she can follow a strict course of treatment', and he suggested Aix-les-Bains on the edge of the Alps.

This argument persuaded Marie-Louise, who was always susceptible to her doctor's advice, and it satisfied Napoleon as well, for he wrote to say that Corvisart's 'noble behaviour more than justifies my confidence in him . . . try to get to Aix, as he advises . . . I will try to write to you from Elba, where I will be making preparations to greet you'. All he had to go on was a faint-hearted letter from Marie-Louise in which she promised to travel to Elba after she had seen her father and taken the cure at Aix, and the argument about the watering-places was to go on well into the summer, providing a convenient excuse for the Emperor Francis to keep Marie-Louise away from her husband without explicitly refusing her permission to join him. It even seems to have convinced Napoleon, although his wife's health was actually good enough for her to make a series of journeys across Europe which were much more tiring than a single visit to Tuscany.

On the Monday evening Marie-Louise at last heard officially from the Austrians, though the message brought by Prince Esterhazy and Prince Wenzel of Lichtenstein came from Metternich rather than directly from her father. Francis had now persuaded the Tsar to abandon his attempt to unite Marie-Louise with her husband, and Shuvalov was told to take the Empress back to Rambouillet, on the road to Paris, and to wait there until the Emperor of Austria arrived to talk to his daughter. They were· equally blunt with Marie-Louise. 'I told them that I couldn't leave without your consent', she wrote to Napoleon that night, 'and they said that they couldn't wait and that even if I wished to leave before seeing my father they had instructions to prevent me doing so. . . . Please don't be angry with me, my dear friend, it's not my fault, and I love you so much that my heart is breaking; I'm so afraid that you may think this is all a plot between my father and myself. . . '

Marie-Louise was now taken away from the rest of the Bonaparte family, who scattered to their several exiles. Before she left Orléans she sent Napoleon a distressed assurance that she had not gone freely. 'I want to be with you,' she wrote; 'the more people try to separate us, the more I long to be near you and look after you.' Her suspicions were justified, for the Provisional Government had given orders that 'at all costs' she must be prevented from taking the road to Fontainebleau. 'The Empress isn't free to follow her own inclinations,' Méneval wrote to Napoleon's secretary Baron Fain next day. 'She is without advice or support against insinuations and force. She is in fact a prisoner—suffering and much troubled.'

Marie-Louise, however, was able to send Napoleon a last and tangible gesture of help. On 12 April, just as she was about to leave for Rambouillet, Napoleon's treasurer Guillaume Peyrusse arrived in search of the Emperor's personal fortune. Marie-Louise told him that she would take half the six million francs that Dudon had left her, partly because she needed money to pay her French cavalry escort and other travelling expenses, partly because she thought it would be safer in her hands than if it was left to Peyrusse to protect it with a handful of National Guards. Peyrusse accepted this suggestion when the Empress gave him a proper receipt and a promise to send the residue on to Fontainebleau, for he was worried enough about the thirty cases of bullion she did give him, and had sent a messenger to Napoleon to say that he could not risk moving 2,580,000 francs in gold through regiments of undisciplined Cossacks. At midnight next day General Pierre Cambronne arrived with one of the crack battalions of the Imperial Guard, after an amazing forced march in bad weather and among hostile troops.

Cambronne was in time to save the money; he was too late to save the Empress as well. By the time he reached Orléans she was well on her way to an enforced and temporary residence in the château at Rambouillet.

★

Caulaincourt and Macdonald returned to Fontainebleau to find the palace almost empty. There were still a few staff and orderly officers on duty, but the imperial household was now reduced to a handful of men: General Bertrand, the grand marshal, and Peyrusse the treasurer, who had both decided to share Napoleon's exile; Maret, who was to remain in France and make himself the leader of the Bonapartist faction; the Emperor's secretary Baron Fain, his medical attendant Dr Yvan, and a score of domestics. It was so quiet, General Pelet remarked, 'that it seemed that His Majesty was already buried'.

The Emperor himself was in a dismal and reflective mood, talking about other great men who had preferred death to dishonour, and saying that the prospect of a humiliating peace made his life unbearable. 'I did everything I could to get myself killed at Arcis,' he said, recalling the last battle in the recent campaign, 'but the bullets wouldn't find me.'

Napoleon was certainly prepared for a soldier's death. He was impassive under fire, and in sixty battles and three hundred lesser engagements no one had ever suspected him of cowardice in the face of the enemy. Yet the excesses of the Revolution had left him with a fear of mobs, and after he became emperor he had grown increasingly uneasy about affronts to his person and the threat of assassination. 'His misgivings on that score', Caulaincourt wrote candidly, 'outran anything that I could imagine', and they fretted him all the time in the last days at Fontainebleau. 'I am surrounded by wolves', he said anxiously to Caulaincourt, and he kept asking whether an English officer would be sent to escort him to Elba, and whether he could make the crossing in an English man-of-war. 'The English people will not tolerate an assassination,' he insisted, paying the first of many such compliments to the honour of a people who had so persistently thwarted his ambitions. 'In that country the ministers have to account for their actions, their undertakings, and even the manner in which they are carried out.'

He found the prospect of self-destruction equally distasteful. 'It would be unworthy of me to think of suicide,' he declared in almost the same breath as he had talked of facing the bullets at Arcis. 'It's the death of a ruined gambler. I am condemned to live.' He made several remarks of the same kind. 'They say a living gudgeon is better than a dead Emperor', he said to Baron de Bausset on 11 May. Yet he was obviously depressed, and the men around him began to fear that he might take his own life: Baron Fain thought that he had been meditating 'a secret design' for several days, his valet Constant later claimed that the palace servants had refused to put a charcoal brazier in the Emperor's room in case he deliberately suffocated himself, and Count Turenne, one of his personal attendants, saw Napoleon toying with his case of pistols and secretly removed the powder from the flask.

All these ominous signs, however, were reported after the peculiar and uncharacteristic episode on the night of 12 April, when it appeared that Napoleon had taken a dose of poison, and it is possible that the anxiety of those around him led them to read their own fears into the event and into their recollections of it. No doubt the Emperor was feeling the bitterness of ruin that night, when Caulaincourt had brought him the text of the treaty and he knew for certain that he had sacrificed his throne without gaining anything for his country, and that his combative impulse was to be confined on a rock in the Tuscan sea, but there was nothing in his response to adversity before or after that one occasion to suggest that he would choose such a final means of escape. A man may be ready to welcome death without seeking it, and Napoleon's capacity for self-dramatization could well have led him in this desperate moment to mimic a hero of antiquity.

He had spent the evening talking to Caulaincourt about his dismal future, and before he retired for the night he wrote a straightforward letter to his wife suggesting that they should travel southwards in company—for he had not definitely heard that Marie-Louise was either unwilling or unable to join him. A few hours later, according to a valet who was disturbed by his movements, he got out of bed, mixed a potion in a glass of water, and drank it. The mixture was reputedly a preparation of opium, belladonna, and white hellebore, said to have been used by the philosopher Condorcet when he poisoned himself to avoid the guillotine, and it was supposed to have been kept in a sachet which the Emperor had carried with him in Russia to cheat the Cossacks if they captured him. A little later, when Napoleon was beginning to show symptoms of restlessness and nausea, the valet fetched Caulaincourt, who sat by his master listening to yet another recital of his feelings of disgrace and his fears that he would not reach Elba alive. It was an odd conversation, in which the Emperor appeared to be playing a premeditated deathbed scene. 'Soon I shall be no more', he said dolefully to Caulaincourt, sending his last wishes to Marie-Louise and his benediction to his son.

There is no means of knowing whether Napoleon had actually taken a dose of poison (staled by age or too weak to be lethal when he had vomited part of it away), whether he had merely taken a sedative and was resisting its effects, or whether he was suffering from nothing except an attack of acute anxiety in which he was seized by a fear of imminent death. Caulaincourt could not at first decide what was wrong, but as he watched Napoleon writhing, hiccuping, retching, and sweating, with spasms of hysteria and moments of calm, he came to believe that he was witnessing an unsuccessful attempt at suicide; and he said that when he eventually called Dr Yvan, who had prepared the sachet in Russia, Napoleon asked the doctor for a strong draught to bring his sufferings to an end. While this request confirmed Caulaincourt's belief that the Emperor had intended to kill himself, it had a most curious effect on Yvan, who cried out in alarm that he was no assassin, dashed from the room without attempting to treat his royal patient, and rode away through the night to Paris.

It was by all accounts a very strange affair, which ended almost as swiftly as it began. Soon after dawn Napoleon seemed to be coming to himself again, and the first sign that he was feeling better was a new round of ranting against the sovereigns who had misunderstood, maligned, and mistreated him. By the end of the morning, after telling Caulaincourt 'I shall live, since death is no more willing to take me on my bed than on the battlefield', he was ready to ratify the treaty; and when Macdonald went in to see him the Emperor seemed pale and weak but quite self-possessed. The fit had passed, and he was soon talking as busily as ever about the new government, about the men who had deserted him when he stood on the threshold of victory, and about the arrangements for his journey to Elba.

He was even beginning to rehearse a new role, in which he would speak like an elder statesman and contemplate the past in the peace of his retirement. 'I cleared the stage of revolutions and healed up the wounds of our own', he said loftily and with some justification in one of his rambling conversations with Caulaincourt.

'People will be surprised to see how resigned I am, and how quietly I propose to live.' He actually spoke calmly about the return of Louis XVIII, saying that France was so well governed that the new king would inherit an excellent administration and capable men to run it. 'If the Bourbons are sensible,' he said, 'they will change nothing except the sheets of my bed.'

★

Napoleon's own affairs were now settled but there remained the problem of Marie-Louise. Soon after the Empress arrived at Rambouillet she received the letter that Napoleon wrote in the morning of 11 April, when he knew for certain that he was to have Elba and that she was to be offered the Parma duchies. If she could not have Tuscany, he said, she should at least ask her father for Lucca and part of the Italian coast facing Elba, and he said he was expecting her to meet him at Briare and travel with him as far as La Spezia.

By now the Empress had realized that she was no longer a free agent, and she was becoming increasingly unsure what she would do if she were. As she waited for her father to visit her the members of her entourage were fighting a boudoir battle for her opinions. Méneval and the Countess de Montesquiou, who had charge of the King of Rome, were urging her to stand by her husband, while Dr Corvisart and the Duchess of Montebello (from personal or venal motives) were doing their best to discredit Napoleon with gossip about his infidelities and his illegitimate children. On 15 April Caulaincourt also drove over to Rambouillet to see what he could do, taking a mistakenly complacent message from Napoleon to his wife. 'Do not insist that she joins me,' he told Caulaincourt. 'I would rather have her at Florence than at Elba if she came there with a frown on her face ... At the Empress's age she still needs toys to play with ... Things can be arranged so that every year I may go and spend a few months with her in Italy, when it becomes apparent that I have resolved to meddle with nothing, and that I am content, like Sancho, with the governing of my island and the pleasures of writing my memoirs.'

Caulaincourt's visit settled nothing. The Empress, he noted, 'received me with great kindness, wept a good deal, and spoke long and feelingly to me about the Emperor', saying in particular that she planned to join him after taking the waters at Aix. But Napoleon remained optimistic, for on the same day that she saw Caulaincourt he wrote her a letter which assumed that she had already seen her father and reached a favourable agreement with him. 'I want you to come to Fontainebleau tomorrow, so that we can leave together', he said, adding that they would be going into exile 'to forget the great things of the world'. As Caulaincourt had realized, Napoleon was always underestimating the hostility of the Allies and the fear which he inspired in them, and this new posture of renunciation was as naive as his persisting belief that his father-in-law would prove to be his friend.

Once again, moreover, delays and distance put Napoleon's timing out of phase. Francis did not actually arrive at Rambouillet until the following day, and by then the conflict of pressures had reduced Marie-Louise to a state of excited indecision. It did not take Francis long to persuade her to return to Vienna 'for a few months', though he was carefully vague about her future when he sent Napoleon a letter written by Metternich over his signature. 'Once having regained her health, my daughter will take possession of her territories,' he said with deceptive ambiguity, 'and this as a matter of course will bring her nearer to Your Majesty's residence.'

Marie-Louise did not think it necessary to dissemble in such diplomatic language about her father's intentions: 'he has been very kind and good to me', she wrote to Napoleon after she had seen Francis, 'but that hasn't eased the terrible shock he gave me, preventing me from joining you, from seeing you, from travelling with you'. It was not only her sense of duty which made her wish for this reunion, she added: 'I really feel that this will kill me, and I pray that you can be happy without me, though it is impossible for me to be happy without you.' Writing to Caulaincourt on 20 April she was even more gloomy at the prospect of going back to Austria. 'What I am afraid of is that they will want to keep me a prisoner there for the rest of my life and will not let me go to Aix', she complained, 'although that is the only way for me to get my health back and to be nearer the Emperor, whom I am most anxious presently to rejoin.' There was no possibility of going directly to Aix, she said. 'I am afraid that I have already promised my father too much to be able to get his permission to do so.'

While Marie-Louise waited at Rambouillet she was visited by the Tsar, who was most agreeable and tried to persuade her that in the end no one could stop her going to Elba though he thought she would be making a mistake if she joined Napoleon in his exile; and she also had a less agreeable call from the King of Prussia, who made 'clumsy attempts to be polite'. But she was not allowed out of the château until a week later when, with twenty-four carriages and a train of baggage-waggons, she set off for Austria, gradually turning a journey which began in distraught misery into something like a holiday tour.

By the time Marie-Louise reached Vienna she seemed glad to be home with her many sisters and the rest of the vast imperial family. But for her father and Metternich her return was not the temporary expedient they pretended. It was a symbolic victory. It showed very clearly that Francis was so opposed to a Bonaparte dynasty that he was prepared to sacrifice his daughter's marriage and his grandson's inheritance to put an end to it. For Francis needed the 'legitimate' Louis XVIII on the throne of France to support his claims on northern Italy as much as Louis, also in the name of legitimacy, would need Austrian assistance to chase Napoleon's brother-in-law Joachim Murat off the throne of Naples. With Napoleon the First out of the way the Habsburgs and the Bourbons had a common interest in ensuring that the King of Rome should never become Napoleon the Second.

3

AN AMBIGUOUS MISSION

On Thursday 14 April an urgent letter from Lord Castlereagh was delivered at the lodging of Colonel Neil Campbell, who had been wounded during the march on Paris and had just arrived in the city, asking if Campbell would 'attend the late chief of the French Government' on his journey into exile, and be ready to leave in a day or two.

Castlereagh had thus reacted positively to Napoleon's feeling of insecurity. He had been unwilling to offer the Emperor sanctuary in England: when Caulaincourt put the proposal for asylum he noted that the Foreign Secretary was 'dumbfounded by my question, and even embarrassed to find an answer', but he had done his best to allay Napoleon's fears. 'He assured me', Caulaincourt wrote later, 'that England (with whom until now there had been no *pourparlers* whatever) would guarantee, in respect of her own actions, the Emperor Napoleon's unmolested sovereignty over Elba, the Parma duchies, etc. . . . Castlereagh was obliging, definite, frank in his sympathies, and most courteous in speaking of the Emperor and discussing his interests; and he kept all his promises to me to the letter.' In particular, the British minister had promised to send an officer to join the escorts provided by the three countries which had actually signed the Treaty of Fontainebleau, and he had chosen Neil Campbell, a well-connected and experienced staff officer who was accustomed to diplomatic missions and to the protocol of European royalty.

In the course of 1813 Campbell had met the King and Queen of Sweden, Crown Prince Bernadotte, the King of Prussia, the Tsar, and a scattering of princes and princesses; and in his present capacity as a military attaché at the Tsar's headquarters he had fought his way across Europe with a Russian corps. On 25 March, in the savage battle at Fère-Champenoise, he had been half killed by a sabre cut in the head and a lance thrust in his back inflicted by rampaging Russian hussars who had mistaken him for a Frenchman, and he was still unfit for duty. His convalescence, indeed, had cut him off from events. 'I had no knowledge of the arrangements in progress regarding the future destiny of Napoleon except through the channel of the daily newspapers', he wrote afterwards. But he was not deterred from accepting Castlereagh's surprising request by his poor health, his ignorance of the Allied settlement with the Emperor, or the loss of all his papers, money, and effects when he was wounded. 'This proposal presented so many points of interest', he said in his dry style, 'that I resolved at all risks to undertake it.'

On Saturday 16 April, at nine in the morning, Campbell called on Castlereagh to receive his instructions, and then went straight to Fontainebleau to join the three other officers who were to conduct Napoleon across France. They were all men of rank. The Austrians had selected General Köller, who had been Schwarzenberg's adjutant in the recent campaign, the Prussians picked Count Truchsess-Waldbourg, who had served Jerome Bonaparte when he was King of Westphalia, and the Tsar detached General Shuvalov from his attendance on Marie-Louise, sending him to Fontainebleau with an order to answer for Napoleon's life with his own.

Castlereagh's formal letter to Colonel Campbell, which was to provide his only authority for the next few months, was less dramatic but equally explicit. All its main clauses dealt with Napoleon's personal safety along the road to the coast and during the short voyage out to Elba. Campbell was to show 'every proper respect and attention to Napoleon, to whose secure asylum in that island it is the wish of His Royal Highness the Prince Regent to afford every facility and protection', to call upon 'any of His Majesty's officers by land or sea' for assistance in his duties, and to stay on Elba indefinitely if Napoleon 'should consider the presence of a British officer can be of use in protecting the island or his person against insult and attack'.

The courtesies due to a defeated enemy were the overt reason why Castlereagh had accepted Napoleon's request for protection, but there was another and covert reason why he had responded to it with such alacrity. He was profoundly worried about Napoleon's security in the opposite sense, and it gave him a chance to install his own confidential agent on Elba and close to the Emperor. Although Britain had no formal obligations in the matter at all, Castlereagh's carefully phrased letter thus made Campbell effectively responsible for supervising Napoleon on the pretext of protecting him; and since the British navy controlled the Mediterranean it was obvious that the means of transporting Napoleon to Elba might also be the means of ensuring that he stayed there. The ambiguity of these arrangements was immediately convenient and ultimately most unfortunate.

Hence the sudden call on Colonel Campbell's services. There were a number of socially smarter and more senior British officers in Paris who could well have been despatched as a temporary escort to the Man of Destiny but it was difficult for Castlereagh to find one who would be willing to stay at Napoleon's side indefinitely, without much reward, and in a very dubious role. Whoever accompanied the Emperor to Elba would have to be something of a guard, something of a gaoler, and something of an ambassador, without enjoying official status in any of these roles. Before Castlereagh wrote to Campbell, in fact, the mission had been declined on these grounds by the elegant Lord Burghersh, who told the Foreign Secretary plainly that he preferred the pleasures of occupied Paris to the doubtful attractions of a barren island off the coast of Tuscany.

Campbell had a very different character. Brought up in a Highland family which regularly sent its sons to be officers in Scottish regiments, he had been trained by years of professional soldiering to accept his duty and to make his living by it. He had served with distinction in the West Indies, had fought under Wellington in Spain and Portugal, had been sent on a special mission to Sweden, and had gone on to spend a year with the Russians as they trudged westwards from Saxony—where he once glimpsed Napoleon across the lines at the battle of Bautzen. At the age of thirty-eight he was a handsome man, with a rather solemn face framed by heavy whiskers, who made a dashing impression as he rode down to Fontainebleau, for he still had a bandage looping out of his thick black hair over his left eye, and a sling for his right arm showed white against the blaze of his red uniform.

He was a reasonably intelligent man, though he did not have much imagination and he was easily bored when he lacked congenial company; he was a competent and industrious soldier—his *Instructions for Light Infantry and Riflemen* was a standard training manual in the British army for many years—and well-mannered as long as he kept his somewhat prickly temper. He spoke French easily and he was a good listener.

All these qualities would help Campbell cope with Napoleon's notorious changes of mood. Sometimes it took the Emperor's fancy to treat his close associates *en soldat*, as one old campaigner to another; sometimes he wanted nothing more than to present himself as a savant or a patron of the useful arts, or to find an audience for his soliloquies on politics and war; sometimes he was simply a Corsican adventurer, bargaining unscrupulously for petty advantages; and sometimes he was a stickler for his imperial dignity. Campbell, moreover, was not merely a brave soldier and an experienced military courtier. He was also a conscientious informant, whose journal and despatches reveal a capacity to draw people out and record in a rather dry and literal-minded fashion what they said and did. It seemed that Castlereagh, for all his haste, had chosen the right man for what was bound to be a very difficult assignment.

★

Only three of the generals who remained at Fontainebleau were to accompany Napoleon into exile, and they were men of strongly contrasting temperament. Cambronne, who left on 14 April with the volunteers of the Guard who were marching across France to embark for Elba, was a rough-and-ready man of combative spirit— 'a desperate, uneducated ruffian' was Campbell's harsh phrase after he had known him for several months. But he was recklessly brave, he had been fighting ever since the first great battle of the revolutionary wars at Jemappes, and he had been wounded so often that another general who went bathing with him in 1813 said that he was 'completely tattooed with scars'. So many men had volunteered to serve under him that he actually marched off with six hundred of the Guard, two hundred more than had been allowed by the treaty, and behind his column of veterans came four cannon, twenty-seven carriages, and a string of Napoleon's favourite horses. There could be no doubts about the loyalty of Cambronne and the *grognards* who made up this praetorian guard.

Nor could there be any doubts about Henri Bertrand. He was an experienced staff officer, dour, obstinate, fussy, and very competent, a born administrator who had found his vocation in managing the imperial household and was unwilling to abandon it even in adversity. He intensely disliked the prospect before him; at the age of forty-one he seemed to be throwing away his career while other men, who had done even better out of the Empire, were scurrying to seek places under the Restoration. His wife Fanny, a handsome woman of Irish descent whose father had been a royalist general before he was guillotined in the Terror, and whose mother was a distant connection of the former Empress Josephine, was even more reluctant than Bertrand himself to exchange the grandeur of the Tuileries for a shabby life of exile on an island that was totally lacking in amenities.

Antoine Drouot was a modest and genial bachelor. Marshal Macdonald described him as 'the most upright and honest man I have ever known, well-educated, brave, devout and simple in his manner'. He had risen to high rank in the Imperial Guard, and when he had to face the painful choice between personal loyalty to the Emperor and a patriotism which would now have to embrace allegiance to the Bourbons he sadly chose to leave France and give up his French citizenship. Unlike Bertrand, who could expect to continue his career in the cramped circumstances of Elba, serving as the Emperor's court chamberlain and chief minister, man of business, confidant, and regular companion (all roles for which there would now be no competition), Drouot expected nothing for himself. He refused a gift of 100,000 francs from Napoleon, saying that it would compromise his reputation for disinterested loyalty and that in any case he had frugal tastes; and when he was offered a post on Elba he at first took it only on the condition that he could resign as soon as Cambronne arrived to replace him. 'I have entirely renounced the great things of the world', he wrote to his old friend General Evain on 5 May, 'and I am going to devote the days of my exile to study.'

★

When Campbell arrived at Fontainebleau he found that General Köller had already made the arrangements for Napoleon's departure on 20 April. All down the route to Lyons the post-houses had been alerted to provide relays of horses, and inspectors had been sent ahead to make sure that these orders were carried out. Now that it had been decided that the Emperor would travel alone, without his wife and son, there was no reason for further delay. Yet Campbell was greeted by a disconsolate Bertrand, who 'expressed himself in most melancholy terms' about the prospects of life on such a *sauvage* and ill-provided place as Elba, asked for the place of sailing to be changed from St Tropez to Piombino, the small port on the mainland just opposite the island, and repeated Caulaincourt's request for Campbell to cross to Elba and remain there as a symbol of British protection.

The reason Bertrand gave for inviting Campbell to Elba (and thereby doing exactly what Castlereagh wanted) was the risk of attack by the marauding Barbary corsairs if the Emperor's party was kept at sea by bad weather, or by the refusal of the French commandant, General Jean-Baptiste Dalesme, to surrender the island. At the end of three demoralizing weeks in which Napoleon had felt 'the pack of wolves' slowly closing in on him, he was so apprehensive that his anxiety about the pirates may well have been as genuine as his fear that royalist agents might abduct or murder him on the road to the south. General Köller, who saw him at Mass early on Sunday morning, noted that he was 'in the most perturbed and distressed state of mind—sometimes rubbing his forehead with his hands, then stuffing part of his fingers in his mouth, and gnawing them in the most agitated and excited manner'.

The other Allied commissioners noted this uneasiness when they were received by Napoleon later in the day. 'I saw before me a short active-looking man, who was rapidly pacing his apartment like some wild animal in his cell,' Campbell noted. 'He was dressed in an old green uniform with gold epaulettes, blue pantaloons and red top-boots, unshaven, uncombed, with the fallen particles of snuff scattered profusely upon his upper lip and breast.' As Campbell was introduced Napoleon smiled, 'evidently endeavouring to conceal his anxiety and agitation by an assumed placidity of manner'; and though General Bonaparte, as Campbell was instructed to call him, for the British did not recognize the imperial titles, was curt and formal with the other commissioners he put himself out to be agreeable to the Scottish colonel. He always timed such talks with the closest calculation. He asked about Campbell's wounds, his medals, his experience of war, and his opinion of the poems of Ossian, the romantic imitations of Gaelic verse which had lately caught the imagination of Europe. He spoke admiringly of the Duke of Wellington, 'a man of energy in war' who on Easter Sunday had won the last battle of the war by defeating Marshal Soult at Toulouse, and he was even more flattering about Wellington's countrymen. 'Yours is the greatest of all nations,' he told Campbell in the terms of grudging admiration which became a commonplace in his interviews with British visitors. 'I esteem it more than any other. I have been your greatest enemy—frankly such; but I am so no longer. I wished likewise to raise the French nation, but my plans have not succeeded. It is all destiny.'

Seemingly much affected, with tears in his eyes, Napoleon paused before coming to matters of detail. Satisfied that Campbell was ready to stay on Elba he went on to ask for a British frigate to sail with the French corvette he had been promised, or, better still, for passage on a British ship, a request Castlereagh granted by return of post, with the added hope that there would be no more excuses for delay.

The commissioners themselves were eager to be on the road for they were in a delicate position. Although there were Austrian troops within call they had no direct means to enforce their will. As long as Napoleon remained at Fontainebleau, with several thousand of his most loyal troops camped in the nearby forests, there was always a risk that he might suddenly become obdurate, or even steal away one night to rouse the units on the other side of the Loire and thus force the Allies to corner him all over again. The four commissioners, moreover, had been given the vaguest instructions, which amounted to little more than an order to escort the Emperor across France, through some districts that were favourable to him and others that were known to be hostile, and to see him safely out of the country. It was a unique task, not least because none of them had any experience of dealing with a man, as Campbell put it, 'whose name had been for so many years the touchstone of my professional and national feelings, and whose appearance had been presented to my imagination in every form that exaggeration and caricature could render impressive'. Even in defeat Napoleon remained an awesome figure, and though diplomacy and battle had at last failed him he could still use charm and his native wiliness to get his way.

Köller, Truchsess-Waldbourg, Shuvalov, and Campbell were well aware of the difficulties that faced them. For a start they agreed to the sharing of information. Up to this moment, for instance, Campbell had not actually seen a copy of the Treaty of Fontainebleau because Britain was not a party to it, and he had no knowledge of its provisions until his colleagues told him. They were, in fact, still holding their first meeting when Bertrand interrupted them with a new set of problems.

'The outside is plain and unprepossessing', Napoleon once said of Bertrand, 'but there is pith and substance inside it.' He was the kind of conscientious worrier whom Napoleon needed to devil for him, and he had now come to complain that General Dupont, the minister of war in the Provisional Government, was violating the treaty by sending an order to General Dalesme in Elba which described Napoleon as 'the former' Emperor (despite the undertaking to respect his title) and by telling Dalesme to evacuate all French troops and property before Napoleon became king of the island. The commissioners said they could not answer for General Dupont, but Köller sent his aide-de-camp Major Clam-Martinitz to Allied headquarters to urge that the French garrison should remain until the men of the Guard reached Elba to replace them, and that Napoleon should be left the stores to start off his military house-keeping on the island.

Bertrand also wanted a change of route. There were two main roads running south to Lyons and Köller had planned to take the most direct line through Auxerre. Napoleon wished to go by the more westerly route through Briare, on the grounds that Cambronne and the men of the Guard were waiting there to see him pass. That route would also take him close to other French units along the Loire; then he could expect to be covered by Augerau's units around Lyons, and to be protected in the last stages of the journey by the Austrian troops who had occupied part of Provence.

The immediate reason for these changes was a report that Louis Bourrienne, the new minister of posts and a former schoolmate and secretary of Napoleon who had been disgraced and had become very hostile to the Emperor, had called in the post-agents stationed along the expected road through Auxerre and given them fresh orders. There may have been a straightforward reason for this move but it made Caulaincourt fear that 'perhaps there was some violence planned in the Midi', or that gangs were being assembled to harass the Emperor. Napoleon's suspicions had already been intensified by other disturbing rumours from Paris over the weekend. It was said that Joseph Fouché, the former chief of police whom Napoleon had dismissed before the fall of the Empire, was back in the capital and busy at his old game of intrigue, telling the Allies that Elba was too temptingly close to the mainland and that they would do better to carry the Emperor off to Russia or to a distant colony. There was also talk of another assassination plot, once more involving the disreputable Maubreuil, which appeared to involve the connivance of some members of the government and the complicity of the Allied commander in Paris; but it came to nothing more than the ambushing of Jerome Bonaparte's wife as she set off for exile in Switzerland.

While Bertrand talked to the commissioners Napoleon sent Caulaincourt off on yet another journey to Paris to complain about these dangerous intrigues and to 'obtain every possible assurance' from the Allied sovereigns. 'The Tsar, exasperated by what I told him,' Caulaincourt noted, 'gave immediate instructions to his minister to take whatever steps were necessary', and even Metternich was comfortingly explicit. 'I do not know whether the decision that has been made is the best one, even for the Emperor Napoleon, let alone us,' he said. 'One thing, however, is certain: the Emperor of Austria has undertaken definite obligations and he will abide by them.'

Caulaincourt had to be content with these promises, though the Emperor was reluctant to leave Fontainebleau before he heard from London that the government there had underwritten Castlereagh's informal assent to the treaty. Almost to the end he seemed to count on a change of fortune in his favour. But he had little to occupy him now everything was settled. He had sent off most of his baggage with Cambronne, but he was still picking over the palace library to find material on Elba and other books which might form the nucleus of a collection on the island. He took a bound set of the *Moniteur*, the official journal which chronicled events from 1789 to 1813, La Fontaine's *Fables*, some volumes of Cervantes and Rousseau, and also Plutarch's *Lives*, which had so often served as his bedside reading. He gave away books, swords, pistols, medals, and coins to officers who came to say goodbye, and sent similar gifts to absent friends and subordinates. With time on his hands he gossiped a good deal, and when there was nobody to distract him he would sometimes stand for a long time on a patch of ground behind the palace, scratching moodily at the gravel with his boot. He had at last come to the point where neither the Allies nor the Provisional Government would tolerate any more delays. His hesitations and equivocations had even begun to try the patience and the generosity of the Tsar.

Almost fifteen years after Napoleon had returned from Egypt, landing at a beach near Fréjus to hurry to Paris and make himself master of France, he was being taken back to that same beach as a ruined man whose only asset was the memory of his former glory.

4

JOURNEY INTO EXILE

Soon after first light on Wednesday 20 April, a cold, grey, and misty day, like many mornings that month, the fifteen carriages which were to convey Napoleon's party across France lined up outside the palace at Fontainebleau. A troop of cavalry was to lead the cavalcade, followed by a carriage containing General Drouot and some other officers who had volunteered to go to Elba, then a *dormeuse de voyage* (a sleeping coach for Napoleon and Bertrand), fifty more cavalrymen, a carriage for each of the Allied commissioners, and nine more for the rest of the household, which included a doctor, an apothecary, a secretary, a steward, two paymasters, six valets, and lackeys, grooms, and farriers to see to the horses. At the same time grenadier companies of the Old Guard filed into the great courtyard of the palace to provide a fitting farewell ceremony for their Emperor.

At nine o'clock, when Napoleon sent Bertrand to assure the anxious commissioners that he would in fact be ready to leave that day, he sat down to write a final note to Marie-Louise. He had sent three the previous day—one to say that the painter Isabey had given him a travelling portrait of the Empress and the King of Rome which gave him great pleasure, another to protest against the Tsar's visit to her and to urge her to bear her troubles 'with steadfastness', and the third to announce that he would be off next morning. Now he sent a last assurance. 'You can always count on the courage, the composure and the kindness of your husband', he wrote, and he added 'a kiss for the little King' as a postscript.

On such a formal and historic occasion, however, Napoleon was not to be hurried, and after some consultation with his staff he sent for General Köller and the other commissioners in turn. He astonished Köller by saying that he was still uncertain whether or not he would leave, and he again complained that the treaty had been violated by the seizure of his private funds at Orléans, by the order to clear French army stores from Elba, and by the refusal of the Austrians to allow his wife and son to join him. Without money or proper protection, he told Köller bluntly, he would not remain on

Napoleon, 1815. An engraving by Sandoz

Napoleon's Arrival on Elba (French cartoon, 1814)

'The Robinson Crusoe of the Island of Elba' (French cartoon, 1814)

'Nap Dreading His Doleful Doom.' (English caricature, April 1814)

Elba; and after talking rather loosely of raising a new army and resuming the war he repeated the suggestion that he might seek refuge in England.

In the middle of this tirade an aide knocked at the door with a reminder from Bertrand that it was eleven o'clock. 'There's something,' Napoleon said testily to Köller. 'Since when have I been ruled by the watch of the grand marshal?' He turned to the aide and said, with teasing bravado, 'perhaps I shan't go after all'. It seemed, Campbell remarked, that the Emperor was 'more and more averse to depart as the time approached', and he went on haranguing Köller in a 'wild and excited style, being at times greatly affected', about the threat that a powerful Russia posed to Austria and about the 'heartlessness' of Francis and Metternich in sending him away without his family. 'If they act with trickery towards me, I will ask for asylum in England,' he said to Campbell. 'Do you think they will receive me?' Napoleon repeated this request so often that Castlereagh was eventually obliged to write on 5 May to get a formal opinion from the Prime Minister; meanwhile Campbell could only give a tactful reply. 'Sire,' he said, 'I presume that the sovereign and the nation will always carry out their commitments faithfully and with generosity.' Napoleon took this remark more positively than Campbell intended. 'Yes,' he replied, 'I feel sure that they will not refuse me'; and after pacing up and down the room he looked hard at Campbell and said 'Very well, we will leave today'. He had a brief word with the Russian and Prussian commissioners, shook hands with Caulaincourt, Maret, Fain and several senior officers and attendants who had stayed to witness his departure, and then he went down the steps into the courtyard to review and address the guard of honour.

It was a solemn moment, and he made the most of it, recalling the victories the French armies had won beneath the imperial eagles, extolling the bravery and devotion of his men, and insisting that even after Marmont and his corps had 'passed over to the enemy' it would still have been possible to continue the struggle against all Europe. But at what cost? 'Foreign and civil war would have devastated our beautiful country', he said. It was to avoid such havoc, he added, that 'I have sacrificed all my rights, and am ready to sacrifice myself, for all my life has been devoted to the happiness and glory of France'. Napoleon was a master of such rhetoric in front of his men, blending dignity with familiarity to unique effect. 'You are all my children,' he declared as he came to his peroration. 'I cannot embrace each one of you, but I embrace you all in the person of your general.' He kissed General Petit on both cheeks, and then turned dramatically to the standard-bearer. 'Bring me the eagles which have led us through so many perils and on so many glorious days.' He covered his face with one of the silk squares, gold-embroidered with battle honours, and was silent for almost a minute. Then, raising his left arm, he spoke again across the crowded courtyard. 'Farewell, my old companions! My good wishes will always be with you! Keep me in your memories!'

As Napoleon walked to his carriage he could hear the sobbing of men who had followed him to the ends of Europe, and now thought they would follow him no more. Those close to him kissed his hands and touched his clothing as he passed, and even the Allied commissioners were affected. 'That was indeed a most moving scene', said Campbell, 'and worthy of the great man who created it.' Afterwards, Captain Pasquin remembered, the old soldiers burnt their flags and divided the ashes among themselves.

It was not quite two o'clock when the Emperor's carriage rolled out of the palace gates, carrying a whole epoch away with him.

★

The fall of Napoleon sent many dignitaries on their travels again. Some were determined to undo what had been done for the rights of man and to restore the rights of royalty. The Tsar and the King of Prussia, having come so far from home, were now going on to celebrate the triumph of legitimacy with the Prince Regent in London. Louis XVIII, whose family had been the symbol of all the old regimes, had not stayed for the festivities in London; he was already making his way back to Paris with a stream of aristocrats and émigrés in his train. The Spanish royal family was returning from detention in France to restore the reactionary Habsburg regime in Madrid. And minor rulers all over Europe were trying to reassemble the pattern of power that Napoleon's armies—and his habit of distributing domains to his relatives and supporters—had so grievously disrupted.

As the victors came in the vanquished moved out. Marie-Louise and her son were leaving for Vienna. Napoleon's mother, Madame Mère, was on her way to Rome with Cardinal Fesch, and though Napoleon was to pass by their halting-place near Roanne he did not turn aside to visit them. His brother Lucien, who had long been a graciously-treated prisoner of war on a country estate near Worcester, was also on his way to the Holy City to seek the protection of a Pope who seemed to bear no grudge for his abduction and long captivity in France; and Joseph, Louis, and Jerome Bonaparte were all in flight towards Switzerland. No one, it seemed, was sorry to see them go. It was only the Emperor himself who was still touched by the magic of monarchy.

He covered a hundred miles on the first day of his journey, stopping at Briare where Cambronne and the *grognards* were waiting to give him a rousing reception and to shout such noisy insults at the Allied commissioners that he had to make a complaint for the sake of appearances. But his private reaction, as he revealed in a letter he sent to Marie-Louise before he left the town, was one of pleasure at these signs of 'attachment and affection'. As he talked to Campbell over breakfast he seemed in excellent spirits, and all through the day he responded to the cries of 'Vive l'Empereur!' along the route, some of them spontaneous, others prompted by outriders sent ahead to warn people that he was on his way. When the party reached Nevers that evening everything appeared to be going well. While they spent the night at Hôtel de la Nation, Major Clam caught up with them, bringing the new orders for the French commandant on Elba, and Campbell, meeting an English officer who had been a prisoner and was on his way to Paris, took the opportunity to send a reassuring despatch to Castlereagh.

The next day, 22 April, started well but ended badly. At Villeneuve-sur-Allier, where the cavalry escort which had come from Fontainebleau turned back, Napoleon felt confident enough to afford a gesture, and with the words 'I need neither my Guard nor the protection of foreigners as I pass through France' he waved away the detachment of Austrian hussars which was waiting to accompany him. A few miles down the road, at Moulins, he began to pay for that moment of pride as a large crowd swarmed about the carriages with shouts of 'Vive le Roi!' and displays of the white cockade. Greatly upset, Napoleon cancelled the planned halt for dinner and hurried on for another sixty miles, reaching Roanne and the cover of an Austrian unit well after midnight.

After the near riot at Moulins Napoleon asked Campbell to ride ahead to make the promised arrangements for a British ship to escort the party to Elba, and before they all left Roanne next morning he amplified this request into a formal demand for transport on a British man-of-war, saying that he might be insulted if he sailed on a French vessel. The first demonstrations of hostility had revived his fears about royalist mobs in the Midi, and he was anxious to get on as quickly as possible, cutting out overnight stops and making everyone snatch a troubled sleep as the carriages rolled on. It was dark when the party reached Lyons, where the Austrian commander was so anxious to avoid friction between the Bourbon and Bonapartist factions that he diverted the convoy round the outskirts of the city, and by noon on Sunday 24 April it was approaching the area north of Valence where Napoleon expected to find Augereau and his army.

Campbell, who was now some way ahead, had been surprised to come across Augereau, who had left his men and was driving north to settle his affairs with the new government. The marshal was disconcerted to learn that Napoleon was coming down the road from Paris. He had always been noted for his extravagantly-worded proclamations and only a week before he had issued a fierce statement to his men, denouncing the Emperor as a tyrant 'who after sacrificing millions to his cruel ambition had not the heart to die like a soldier'. He now repeated this accusation of cowardice to Campbell. The Emperor, he said, 'ought to have marched full upon a battery and thus put an end to himself'.

Although Augereau told Campbell that he would repeat this charge to Napoleon in person he was clearly embarrassed when he met the Emperor later that day. It was a difficult encounter for them both. Although Napoleon was almost certainly aware of Augereau's insulting manifesto, he kept himself under control, treating Augereau much as he had treated the other marshals who had abandoned him for the sake of peace, and merely reproving him for his reluctance to deploy the reserve army in the last stages of the war.

In the eyes of Augereau's soldiers, indeed, it was the marshal who had been the faint-heart, and guilty of treason in forcing them to wear the white cockade, and as Napoleon passed through some of these units the men lined the road to cheer him and one junior officer after another came up to him with pledges of loyalty. 'If you had twenty thousand men like me', one cavalryman said to him as he rode into Valence, 'we would release you and set you at our head again. It is not your soldiers who have betrayed you but your generals.'

★

It was a different story at Montelimar, where the sub-prefect greeted Napoleon with the ominous news that royalist agents had ridden ahead to raise a mob against him at Avignon, and that the occupation of Marseilles by the English had sent royalist demonstrators careering wildly through the city. Napoleon considered, then rejected the idea of a change of route, and went on through the darkness in the hope that the hostile crowds would have dispersed before he arrived at Avignon in the small hours of the morning. At Donzère, which he reached at nine in the evening, an angry throng almost blocked the way out of the little town, and in every village people were waiting to see him go by and to shout at him.

He had some reason to fear an ambush. The district had been roused by Colonel Mollot, one of several envoys whom the Comte d'Artois had despatched across France to rally support for a Bourbon restoration, and Mollot had already whipped up the people of Avignon to bar the road to some carriages which Napoleon had sent ahead of the main party. By the time Napoleon himself arrived the demonstrators had been up for most of the night and they were beyond the control of the guard which the mayor had hastily improvised. As they surged about the line of vehicles, chanting and throwing stones, Mollot approached Napoleon's own carriage waving a sword; he was held off by the combined efforts of Bertrand and Drouot, and the groom threatened him with a pistol while the coachman lashed at the horses and drove Napoleon away on a detour round the walls.

The reception at Orgon, twenty miles down the road to Aix-en-Provence, was even more alarming. The hostile inhabitants had hung a blood-smeared effigy of Napoleon on a makeshift gallows outside the post-house, placarded with the ominous slogan 'Such will sooner or later be the fate of the Tyrant!'; and while the ostlers struggled to change the horses people threw stones at the carriage windows and jeered at Napoleon. He sat silent, sheltering behind Bertrand, but the situation was becoming so disagreeable that the Allied commissioners and their aides stood between Napoleon's coach and the crowd, and General Shuvalov made an attempt to calm the mob with a speech. 'It is shameful to insult a defenceless unfortunate, a person who saw himself as the ruler of the world and is now at your mercy,' he declared. 'Leave him alone. Look at him, and you will see that contempt is the only weapon you can use against a man who is no longer dangerous.'

Napoleon was badly shaken, and he became even more fearful when he was stopped about a mile beyond Orgon by a man who passed on a new rumour that Talleyrand had sent assassins to catch him before he reached the sea. At St Cannat, where there was another change of horses, he insisted on leaving his carriage and dressing in a plain blue coat, even taking the humiliating precaution of placing a white cockade in his hat. Then, disguised as a courier, he rode ahead with one of the grooms, across the rocky landscape and into the mistral which was lifting the dust. He rode straight past the next post-house (where a waiting crowd angrily attacked his coach and Bertrand had to be extricated by the commissioners) and he stopped at a wayside inn at La Calade. When the rest of the party caught up with him he was closeted in a small room, with his head in his hands and tears in his eyes. He had been quite unnerved by the landlady who, at the mention of Bonaparte's name, started to denounce the Emperor for the death of her son and nephew. 'I want to see whether he will manage to escape,' she said, unaware that he sat before her. 'I think the people will murder him, and there is no doubt that the scoundrel deserves it.'

The whole party was exhausted and hungry, but Napoleon could neither rest nor eat for fear that he would be trapped or poisoned. 'At the least noise', Truchsess-Waldbourg noted, 'he started up in terror and changed colour.' He was alarmed to discover that there was a grill across the window, and though politeness made him sample every dish 'he spat everything out on to his plate or behind his chair'. He would take nothing except a morsel of bread and a bottle of wine brought from his carriage. 'He allowed himself to be completely overmastered by his fears,' Major Clam wrote afterwards. 'He was white and disfigured, his voice was broken, he could not manage to remain calm even before the domestics.'

General Köller sent a similar report to Metternich, speaking of 'the strange and trying hours we passed at the inn, when the Emperor was overcome with anxiety, thought of nothing but the means to be adopted for his safety, the disguise he should adopt, etc.' Worried by the mood of the crowd that was gathering outside the inn, and by rumours of an uglier reception in store at Aix, Napoleon suggested that the party should go back to Lyons and try another route to Italy. The commissioners refused, and Napoleon declined their alternative suggestion that they should go straight to Toulon, well known for its royalist sentiments. He then asked if he could pass as Colonel Campbell, or some other British officer; in the end, before the party left at midnight, he was kitted out in an Austrian uniform donated by Köller, a Prussian forage-cap lent by Truchsess-Waldbourg, and a Russian cloak belonging to Shuvalov, while Shuvalov's aide-de-camp put on the courier's costume which Napoleon had worn earlier in the day.

Even as they all left the inn, he was still thinking of a subterfuge to protect himself from attack. For he travelled with General Köller, and to give the impression that the passenger was no one of importance he told the coachman to smoke and asked Köller to sing some kind of ditty. When Köller protested that he had no voice Napoleon begged him to whistle a tune. 'And with this singular music we made our entry into every place', Truchsess-Waldbourg remarked, 'whilst the Emperor, fumigated with the incense of the tobacco-pipe, pressed himself into a corner of the carriage and pretended to be fast asleep.'

In six days the party had come five hundred miles from Fontainebleau, and it had been a trying and disagreeable journey. Napoleon, indeed, was well aware that his fears had made a bad impression on the Allied officers who had witnessed his collapse. 'I have shown myself to you quite naked', he confessed to Köller, and as his spirits began to revive next day he began to assume an imperious manner, as if to efface the unpleasant memory. In any case, once he passed Aix and St Maximin he felt safer, for the local population was less antagonistic, and Shuvalov had sent messengers ahead to turn out the gendarmes and alert the Austrian units along the route.

★

By the evening of Tuesday 26 April the Emperor had reached Bouilledou, a large farmhouse near Le Luc, where he was to meet his favourite sister Pauline and stay the night. No one else in the family for whom he had done so much had made any effort to see him after his defeat, let alone offered to share his exile, but Pauline was always loyal in her spoilt and self-centred fashion. On 21 April she wrote to her brother-in-law Bacciochi, who had fled from Tuscany to Rome, to say that she expected Napoleon to pass through the part of Provence where she was staying on her way back from a tiny spa in the Maritime Alps, and that she hoped to go with him to Elba or on to Naples to join her sister Caroline and her other brother-in-law Joachim Murat. 'I have not loved the Emperor as my sovereign,' she wrote; 'I have loved him as my brother, and I shall always remain faithful to him until I die.'

The Princess Pauline Borghese was more sensible than her sister Elisa, a duller libertine, and more lively and charming than her younger sister Caroline, a born intriguer who devoted herself to the difficult task of keeping her impulsive husband on the throne of Naples. At thirty-four she was still a lovely woman. The great Italian sculptor Canova had twice modelled her as Venus, and she had taken full advantage of her charms in her many affairs. Her first husband, General Leclerc, had died of yellow fever in the West Indies. Her second, Prince Borghese, was a handsome dilettante who came to tolerate her infidelities. Napoleon was always trying to control her behaviour and limit her extravagances, but for the most part she ignored his moral homilies and his fussiness about scandals in the Bonaparte family, counting on the fact that he had always been devoted to her. His enemies, in fact, liked to circulate anecdotes which suggested that there was something more than a natural affection between them.

Although Pauline had a reputation as a hypochondriac, given to fits of indisposition, she really was a sick woman, suffering from salpingitis, a uterine infection which caused recurring and debilitating discomfort. She was sufficiently unwell when Napoleon arrived to plead that she could not immediately accompany him to Elba— though she was actually waiting for one of her lovers to settle some of her complicated financial affairs in Paris. She promised, however, that before long she would call at Elba on her way to Naples, and it is probable that the question of Murat's future was one reason for Napoleon's visit. Pauline of all people could most easily carry a confidential message from the Emperor to the King of Naples.

The brothers-in-law had been at loggerheads ever since Murat defected in 1814 and made a hasty alliance with the Austrians. Now they were to be close neighbours, and they would both have to be very circumspect in the coming months. On the one hand, Napoleon would certainly be suspected of intriguing with Murat whether he did so or not; and he would therefore be as keen to avoid giving the Allies an excuse to whisk him away from Elba as Murat would be eager to avoid provoking them into throwing him over and restoring Ferdinand I, the Spanish Bourbon whom the British were protecting in Sicily. On the other hand both Napoleon and Murat were perfectly well aware that the moment might come when they would need each other again, and that they would be wise to keep covertly in touch while they maintained a public posture of disagreement.

★

Campbell had now made all the arrangements for the final stage of the journey. He reached Marseilles in the evening of Monday 25 April, made contact with Captain Thomas Ussher who had taken possession of the town on the previous day when he sailed in with the frigates *Undaunted* and *Euryalus*, and used Castlereagh's sweeping authority to send the *Undaunted* on to St Tropez. There Ussher found Captain de Montcabrie, who was standing by with the frigate *Dryade*, the corvette *Victorieuse*, and a transport ship, and expecting to ferry Napoleon's party over to Elba.

When Campbell hurried back overland to join the other commissioners at Le Luc he found that there had been a change of plan. Napoleon now wished to spend his last night in France at the Chapeau Rouge, the inn at Fréjus where he had stayed on his return from Egypt, and then to leave next day from the beach at nearby St Raphael. Campbell therefore sent orders for Captain Ussher to sail across the bay to the new rendezvous, which he reached at noon next day. As soon as the *Undaunted* dropped anchor Napoleon again made it clear that he preferred to embark in the British frigate than to cross to Elba under the Bourbon flag, and when the French vessels arrived later in the day they found that Ussher's working parties were already loading the Emperor's baggage, carriages, and horses. Captain de Montcabrie offered to dispense with the royal colours if Napoleon would change his mind, but when Napoleon realized that the *Dryade* was merely to serve as an escort, and that he was to embark on the much smaller *Victorieuse*—which he was then to keep in Elba—he confirmed his decision to sail with Ussher. Both the French ships sailed away next day.

Captain Ussher, meanwhile, had settled an important point of naval protocol about Napoleon's reception. 'I was desirous to treat him with that generosity towards a fallen enemy which is ever congenial to the spirit and feelings of Englishmen', Ussher wrote. But he was very doubtful about the propriety of greeting him with a royal salute of twenty-one guns, as Napoleon expressly desired. He at first tried to evade the point by suggesting that Napoleon should go on board after dark, as the British navy did not fire any salutes after sunset. When that expedient was rejected, Ussher turned to the commissioners, who assured him that the Treaty of Fontainebleau described Napoleon as 'Emperor and Sovereign of the Island of Elba'; and he was finally satisfied when Campbell showed him Castlereagh's letter of instructions.

The commissioners themselves had business to settle before

Shuvalov and Truchsess-Waldbourg set off for Paris and left Köller and Campbell to go on to Elba, for on their arrival at Fréjus Bertrand had sent Campbell a long note setting out points of protocol and convenience on which Napoleon wished to be satisfied before he would sail. Campbell was far from happy with these requests, 'founded upon a treaty which had never been *formally* exhibited to me', and he had no idea how Castlereagh would react if he exercised his discretion to make arrangements to carry Pauline and the Guard to Elba in British ships, or to make sure that Napoleon had all the 'domestic necessities' he needed. He covered himself by consulting the other commissioners and securing their signatures to a letter of agreement. But he was rapidly moving into a situation where there would be no one to consult, and he would have to act as best he could, without instructions, authority, or any clear idea what Castlereagh or Britain's allies expected him to do.

As soon as Napoleon was given these undertakings he drafted the letter which Drouot was to deliver to General Dalesme when the party reached Elba. Once more he asserted his right to the stocks of munitions and stores on the island, and for the first time he mentioned the subjects of his new and very small 'imperial domain'. Tell them, he wrote to the French commandant, that 'I have chosen their island as my abode on account of their pleasant manners and the excellence of their climate'. And with that letter his official business in France was finished.

'I was very satisfied with the commissioners', Napoleon wrote in a short letter to Marie-Louise that morning, and in the evening he gave them a farewell dinner. Captain Ussher was also a guest, and he noted that while Napoleon talked 'with great animation' about naval matters, 'he appeared to feel his fallen state'. After the party Ussher spent the night on shore, and he was woken at four in the morning by two agitated citizens who said that groups of men had broken away from Eugene's army and were marching into France to free the Emperor. The commissioners, Ussher remarked, 'were as little pleased as I was at being awakened at so unreasonable an hour', but they were all uneasy, and their doubts were increased when it became 'pretty evident that Napoleon was in no hurry to quit the shores of France, and appeared to have some motive for remaining'. Ussher then went to Napoleon, pointing out that he expected a change of wind to the south which would oblige him to leave the lee shore on which he was anchored. After delivering what amounted to an ultimatum that Napoleon must soon sail or take the consequences Ussher went out to his ship to prepare for the Emperor's reception, and at ten in the morning a boat put out with a message from Campbell saying that Napoleon felt unwell but hoped to leave within a few hours. He saw, Ussher drily observed, 'the necessity of yielding to circumstances'.

During the day Napoleon again wrote to Marie-Louise, saying little beyond a conventional farewell from 'an affectionate and faithful husband', and enclosing a copy of a letter he had just addressed to her father. 'The desire of the Empress and myself is to be reunited,' he said, 'especially at a time when we have all felt the rigours of fortune. Your Majesty is of the opinion that the Empress needs to take the waters, and that immediately afterwards she will come to Italy. This is for me a pleasant prospect, and I count upon it.' Bertrand wrote in a similar vein to Méneval. 'You must be aware how greatly we desire that the Empress should divide her time between Parma and the island of Elba,' he said, making it clear that the Emperor's expectations had now been scaled down; 'we should be so happy to see her now and then.'

Napoleon was at last ready to leave, and about eight in the evening Ussher came in to escort him to the ship. A small and curious crowd was waiting in the street, and Ussher tactlessly remarked 'that a French mob was the worst of all mobs'. Napoleon, who seemed 'a good deal agitated', concurred. 'Yes,' he said, 'They are like a weathercock.'

It was a bright and moonlit night, with little wind. 'The road was lined with Hussars, and Square was formed on the Beach round the Boat,' Lieutenant Hastings of the *Undaunted* wrote to his family. 'This mighty enemy of England prefer'd trusting himself in the hands of those very people whom he had so often stigmatized as being destitute of Honor and Principle, to those over whom he had reigned, and so often led to victory and glory.' It was, as Hastings observed, a paradoxical scene. 'Deserted by all his Generals but two as well as by the greater part of his Domesticks; and ever fearing for his own safety, he throws himself on board a Frigate belonging to that country, whose most deadly enmity he justly merited.'

A bugle sounded the honours. Napoleon said goodbye to Shuvalov and Truchsess-Waldbourg, and crossed a small plank into Ussher's barge. As the seamen pulled the boat up to the *Undaunted*, lying close-to and ready to sail with the morning breeze, the first cannon fired in the roll of the royal salute.

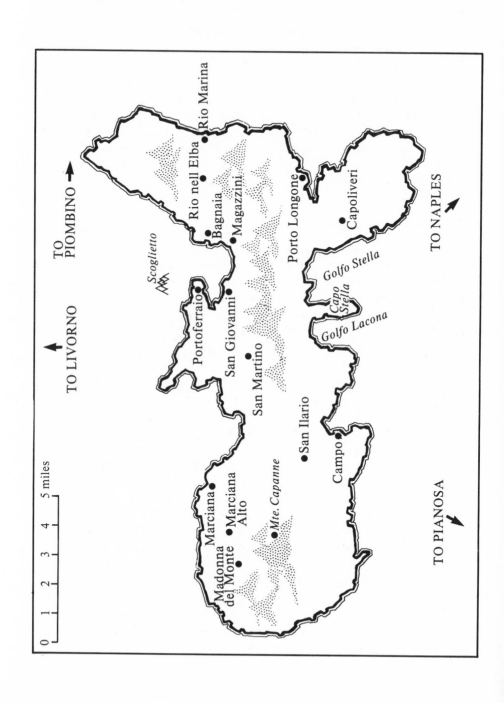

TO LIVORNO

TO PIOMBINO

TO NAPLES

TO PIANOSA

0 1 2 3 4 5 miles

Marciana
Madonna
del Monte
Marciana
Alto
Mte. Capanne

Portoferraio
Scoglietto
San Giovanni
San Martino

San Ilario
Campo

Rio Marina
Rio nell Elba
Bagnaia
Magazzini

Porto Longone
Capoliveri

Golfo Stella
Capo
Stella
Golfo Lacona

5

A CHANGE OF FORTUNES

Captain Thomas Ussher came from a notable Anglo-Irish family. His father was the first Astronomer-Royal of Ireland, and another of his forebears was the famous archbishop whose dating of the Creation at 4004 BC was an article of fundamentalist faith well into the nineteenth century. Ussher himself, *The Times* later remarked, was 'a gentleman of kind and excellent heart', and after serving all through the wars against Napoleon he was fascinated by his famous passenger. Only a few months before, indeed, when he had received some English newspapers which speculated on the Emperor's defeat and the possibility that he might try to escape to America, Ussher had cut out a detailed description of Napoleon from the *Courier* and put it up in his cabin, jokingly telling dinner guests from other ships that 'they had better take a copy of it, as he might possibly come our way'.

Now the jest had come true. 'It has fallen to my extraordinary lot to be the gaoler of the instrument of the misery Europe has so long endured', Ussher wrote home on 1 May, reporting this 'glorious finish' to his career and adding that he found the Emperor so crestfallen that 'I endeavour by my attentions to quiet his uneasy mind'.

It was easier to be agreeable to a defeated emperor than to fit the remnants of an imperial court into the cramped quarters of a British frigate. Ussher gave Napoleon the roomy captain's cabin which ran across the stern of the *Undaunted*, installed Drouot and Bertrand in his night-cabin, and by moving out the other officers and the midshipmen he found berths for Campbell, Köller, and Clam, for Colonel Jerzmanowski, who was to command a detachment of Polish lancers as part of Napoleon's guard, for Peyrusse, two gentlemen who were to be Grooms of the Bedchamber, the Controller of the Household, a physician, an apothecary, and a secretary. He also had to accommodate twelve minor officials and ten servants.

Foul winds made the cramped conditions worse, forcing the *Undaunted* to take five days for a crossing that was normally made in two, and the heavy seas caused Köller, Drouot, and some other members of the party considerable misery. But Napoleon was a good sailor, and Ussher noted that 'his spirits seemed to revive suddenly' once he was on board an English ship. He spent much of his time on deck, watching the crew at work, addressing his characteristic rapid-fire questions to the officers, and displaying a good knowledge of ship-handling and navigation. This display of *bonhomie* did not please everyone. Lieutenant Hastings, closely watching the man he had been brought up to regard as a tyrant, wrote home that Napoleon was physically 'inactive and unwieldy', that his countenance was 'by no means agreeable', that his manners were 'far from dignified or graceful', and that he 'assumed an affability which certainly did not appear to be natural to him'. Ussher was much more susceptible to Napoleon's flattery. At the end of the voyage he felt that he had carried out his 'difficult mission' with 'that deference and respect for the feelings of Napoleon which have appeared to me no less due to his misfortunes than to his exalted station and his splendid talents'. Other officers, surprised that so great a personage should be so cordial and approachable, were equally impressed; and the members of the crew, whose curiosity about 'Boney' turned to gratitude when he gave each man a parting present of a bottle of wine and a gold coin, roared approval as their boatswain expressed their thanks and hoped that the defeated ruler of Europe would enjoy good health, a long life, 'and better luck next time!'

Captain Ussher and the officer of the watch dined with Napoleon, Bertrand, Drouot, Campbell, Köller, and Clam, and though Ussher spoke so little French that he had to rely on Campbell's translation, he was spellbound by the ease and range of talk at his table. Naturally inquisitive, well-read, well-informed, endowed with an excellent memory, Napoleon knew how to make the most of such occasions, and as he sat in the after-cabin of the *Undaunted* he gossiped almost as one officer to another about old triumphs and defeats.

Yet in between the talk about Trafalgar, and the abortive plan for the invasion of England, there were also some worrying hints for the future. At breakfast one day Napoleon spoke patronizingly about the returning royalists as 'poor devils', though he quickly corrected himself to describe them as great lords who simply wanted to recover their estates and their privileges; and he went on to predict that if they allowed the English to impose a commercially damaging peace on France 'the Bourbons will be driven out in six months'.

Campbell was alert for any such sign that Napoleon was still reluctant to accept his fate, and when the *Undaunted* met the English brig *Merope* on 29 April he took the chance to send cautionary messages to Admiral Sir Edward Pellew, the naval commander in the Mediterranean, and to Lord William Bentinck who was in charge of the British forces protecting Ferdinand I in Sicily and had lately occupied Genoa. Two days later, running for shelter in the lee of Corsica, the *Undaunted* fell in with the *Berwick*, the *Aigle*, and the *Alcmene*, which were escorting a convoy of troops to Ajaccio, the Corsican town where Napoleon was born in 1769. Campbell at once sent off fresh letters to Pellew and Bentinck in Genoa, and a despatch to Lord Castlereagh in Paris, in which he reported Napoleon's 'indiscretion' about the Bourbons; and though he allowed Napoleon to write some private notes for delivery in Corsica he was sufficiently suspicious to pass them over to the commander of the British force with the request 'that they might either be destroyed or opened'.

Campbell was clearly uneasy, for when he wrote to Castlereagh on 1 May he asked explicitly for fresh instructions. By 3 May, when the crossing was over, he was even more anxious and he drafted another despatch to Castlereagh summarizing his impressions of Napoleon. 'He possesses no command over himself in conversation', Campbell wrote, 'and although at Fontainebleau he expressed his desire to pass the remainder of his life at Elba in studying the arts and sciences, he has given frequent proof of the restlessness of his disposition, and his expectation of opportunities arising which will give scope to the exercise of it in France. He persuades himself that the greater part of the population in France is favourable to him, excepting in the neighbourhood of the coast.'

General Köller also thought the Emperor's old habit of intrigue died hard, noting that Napoleon harped on the possibility of disagreement between the victorious Allies. More than once, Köller reported to his British colleague, Napoleon had tried to persuade him of 'the apprehension which ought to be entertained by the Emperor of Austria and Prince Metternich in consequence of the increasing power of Russia'.

By the morning of 3 May the weather had cleared and the *Undaunted* was making better progress towards Elba. It was still not certain, however, that Ussher would be able to land his passengers. As the *Undaunted* came abreast of the island of Capraia, a small fishing community to the north of Elba, a party boarded the frigate to report that the islanders had overthrown the French garrison and wanted British protection. Elba, it seemed, was also in a state of turmoil, and no one knew for certain who was in control. After landing one of his officers on Capraia to serve as a temporary authority

Ussher stood on southwards in a breeze so light that Napoleon became 'exceedingly impatient'; he paced up and down the foredeck, and as soon as the masthead lookout sighted the forts at Portoferraio, the main harbour and capital of Elba, he was agog to know what flag was flying over them. This was also a matter of some moment for Captain Ussher. It would be very embarrassing to find that Elban rebels had taken over the island in the hope of a British occupation, and as a last gesture of defiance against Napoleon, for he had come to install Napoleon himself in accordance with the Treaty of Fontainebleau. It would be equally embarrassing if General Dalesme had not yet received news of that treaty or of the return of the Bourbons and so stood firm under the tricolour; 'if we had had a fair wind', Ussher wrote afterwards as he reflected on the quirks of history, 'I should have found the island in the hands of the enemy, and consequently must have taken my charge to the commander-in-chief, who would, no doubt, have ordered us to England'. Fortunately for Ussher fresh orders had already reached Dalesme; the white flag of the Bourbons now flew from Fort Falcone and Fort Stella, whose strong ramparts hooked round Portoferraio like a shielding arm, and Ussher knew that his mission was safely completed. As the failing wind had left the *Undaunted* becalmed in the afternoon sun he sent Jerzmanowski, Drouot, and Clam ashore to present General Dalesme with the official notices of Napoleon's arrival and to make arrangements for the official reception next morning. Campbell and Lieutenant Hastings went with the official emissaries to assert the British presence.

As the officers were being rowed across the harbour Napoleon remembered that 3 May was the twenty-fifth anniversary of the procession of the States-General through Paris at the beginning of the French Revolution. And on this very day, as Napoleon looked up at the bleak mountains which hemmed his tiny capital against the sea, Louis XVIII reached St Denis, entered Paris in a coach decorated with the royal lilies, and drove straight to a celebratory mass at Notre-Dame.

★

Monte Capanne is the highest point on Elba, rising steeply to 3,300 feet above the Mediterranean, and the view from its summit is spectacular. To the east is Italy, a strip of blue coastline between Livorno and Civita Vecchia, and at the nearest point only six miles away across the Piombino channel. To the west is the long shadow of Corsica, Napoleon's home until 1792, when the Bonaparte family sided with the French against the Corsican separatists and left to seek a new life in revolutionary France. In between Italy and Corsica, and running north and south from Elba, lies the chain of islands which form the Tuscan Archipelago. The most northerly is Gorgona, the 'escarped rock' where the youthful Napoleon, aspiring to be a novelist, set a romantic tale of friendship between an exiled Corsican patriot and a freedom-loving Englishman. Capraia, a little larger, was a rough place in Napoleon's time, and its people lived by keeping goats and fishing. Next come Elba and its attendant rocks, of which only Cerboli and Palmaiola are big enough to take even a watch-tower; then to the south-east, Monte Cristo, a bare cone that reaches 2,000 feet; and, to the south-west, Pianosa, a flat piece of land so isolated and vulnerable to attack that for long periods between Roman and modern times it was left unpopulated.

Napoleon was not even given all this small set of islands. He could be sure of nothing except Elba, and that looked far from promising. Everything that he could see from the deck of the *Undaunted*, indeed, seemed to confirm what he had read in *Voyage à l'île d'Elbe*, a gloomy account published by Thiébaut de Berneaud in 1808 which he had found in the library at Fontainebleau and used as his main source of information on his new domain. 'The mountains of Elba', Thiébaut wrote, 'together present only a mass of arid hills which fatigue the senses and impart sensations of sorrow to the soul. Rough and uneven roads, deserted cottages, ruins scattered over the countryside, wretched hamlets, two mean villages and one fortress— these, generally speaking, are all that meet the sight on the side of the island which runs along the channel of Piombino.'

The remainder of Elba was much the same. Sixteen miles long, seldom more than seven miles in width, with a rocky and largely inaccessible shoreline, the key-shaped island was little more than two mountain ranges joined by a lower neck of land. Rich in rocks, especially marble, granite, and the iron ore which had been mined for centuries, it was poor in almost everything else. The rough hillsides, covered in box, myrtle, and tamarisk, were useless for anything but goats. There were some vines, a few olives, groves of sweet chestnuts that were pounded into a coarse flour, but there was so little flat land in the waist of the island that the islanders could grow only a third of the grain they needed; the rest of the grain, as well as meat, dried vegetables, and any manufactured goods had to be imported, and the island's trade was precariously balanced by the earnings from the iron mines, the salt-pans, and the seasonal tunny fishery which were the only profitable and taxable assets.

It was a sparse place, sparsely populated. When Napoleon arrived there were only 12,000 people on Elba, and a quarter of these lived in Portoferraio, in a maze of houses which rose in steep tiers between the waterfront and the forts above. It was an attractive site, for the town was built on a peninsula and it turned its back to the sea to face southwards across the harbour. But for all that the capital was no more than a small and seedy Mediterranean port. No one was wealthy and most people were poor, and the town was so unattractive to visitors that it managed with only one grubby inn. Since there was no regular supply of water, and the inhabitants threw their ordure into the fetid alleys, it was also dangerously insanitary: there were many cases of dysentery, typhus, malaria, and leprosy, and a poor diet was the cause of widespread scurvy. In the villages and the countryside life was even harder. Families lived in wretched huts, slept naked on one bed, drank rough wine flavoured with ginger to make it palatable, cooked in crude earthenware pots, and considered themselves lucky when there was salt meat or fish to supplement the staple diet of goat cheese, coarse bread, and chestnut meal.

The Elban people, in fact, were very like the Corsicans whom Napoleon had known in his youth, and they lived the same sort of rough and independent life. Although Thiébaut remarked that they were 'extremely irritable, and impatient of contradiction', and thought them 'almost universally ignorant and credulous', they had vigour, pride, and an engaging taste for spoken poetry. They were also very tough, for they were the survivors of a cruel history. The natural hazards were bad enough—poor land, droughts, sudden storms that ruined crops and drowned fishermen; but the human afflictions were even worse. For more than two thousand years the island had been conquered, exploited, raided, pillaged, bombarded, and blockaded, its people had been killed, ravished, and carried off as slaves; and some of its villages were so thoroughly destroyed that not even their ruins remained. One abiding sign of this insecurity was the duplication of coastal villages higher up the mountain. When the raiders struck from the sea the people of Marciana could retreat to Marciana Alto, from Campo to San Ilario, from Rio Marina to Rio nell Elba.

It was an island of legends. It was said that Jason anchored the

Argo in the northern bay after he had captured the Golden Fleece, that Aeneas raised levies in Elba, that Elbans fought in the defence of Troy against the Greeks, and that St Paul preached there. It was also an island that had known little peace or settled government. The Etruscans held it, and mined its iron; the Phoenicians traded there; the Romans made it a colony. After the fall of Rome it was thoroughly sacked by the barbarians. The Saracens came, again and again, and when the Pisans drove them off in the eleventh century the Pope gave the island to the Republic of Pisa as a reward, and a Pisan tower still stands on the harbour at Marciana. It was captured by Genoa, sold to Lucca, racked by the factional wars between the Guelphs and Ghibellines, claimed by the Spaniards in the seventeenth century, twice devastated in raids by the Turkish corsair Barbarossa, and then given by Philip II of Spain to Cosimo de Medici in part payment of a debt. Cosimo had ambitious plans, setting out to establish a new order of chivalry called the Knights of St Stephen, and planning to make Cosmopolis—a town on the site of modern Portoferraio—the rival of the fortress that the Knights of St John had built in Malta. He commissioned Gianbattista Belluci to design the great forts and the city walls, and sent over workmen from Florence to begin them, but the scheme was never finished. The Spaniards came back, setting up another fortress at Porto Longone (the modern Porto Azzuro on the east coast) while the Florentines kept Portoferraio, and the Principality of Pisa had a shadowy hold on the rest of Elba.

In the eighteenth century the Spaniards, Germans, and French fought for the island, and on it. In 1794 when the British brought off the royalists from Toulon they settled them in Elba, and held it for three years. Then Napoleon conquered Italy, the Royal Navy was temporarily forced out of the central Mediterranean, and the British evacuated the island. In 1799 the Elbans massacred the French troops which occupied the island, yet after the Peace of Amiens in 1802 it briefly became part of metropolitan France, sending deputies to the National Assembly in Paris. Then, in 1805, when Napoleon made his sister Elisa Bacciochi the Princess of Lucca and Piombino, he gave her Elba as part of her endowment. A place which had changed hands so often, which lacked any certain allegiance or military importance, and had so lately been ruled distantly by Napoleon and more directly by his sister, was an appropriate estate for a man who had so often redrawn the frontiers of Europe, and created kingdoms, principalities, and dukedoms on the whim of a moment. Now the Tsar had done the same for him, and he had to make the best of it.

★

Napoleon's arrival on the *Undaunted* was a relief for General Dalesme, who was far from sure what had been happening outside Elba and had almost lost control of the island itself. He was supposed to command a garrison of 5,000 men, but it was a weak and disaffected force, filled out with reluctant levies from Italy and Corsica; and while it sufficed to make a show of strength in the fortifications at Porto Longone and Portoferraio, it was no longer able to keep order in the rest of the island, and on 21 April some of the troops mutinied at Porto Longone. Dalesme could do nothing but repatriate the Italians, let the deserters run free, and withdraw the 500 trustworthy French soldiers of the 35th regiment into the forts at Portoferraio. On the south side of the island the villages of Capoliveri and Campo had declared for Murat and the kingdom of Naples. At Marciana, where the townspeople burned an effigy of Napoleon, they told the captain of a British ship that they hoped for a British landing, and crude copies of the Union Jack were hoisted in several other villages. The Imperial regime had collapsed in Elba just as the Allies were despatching the former Emperor to govern it in person.

This astonishing and paradoxical news had reached Dalesme in such odd ways that he was most suspicious. On 27 April a British frigate and brig had made contact with the insurgents at Marciana and then sailed along the coast at Portoferraio, where the captain of the frigate sent a boat ashore under a flag of truce. The messenger brought some French newspapers which reported Napoleon's abdication, and also a demand for the surrender of Elba on the grounds that a peace treaty between France and the Allies had ceded the island to Britain. It was an ingenious initiative by the commanding officer of the frigate, but it failed. After defying the blockading British navy for so long Dalesme was not to be bluffed into a last-minute capitulation, and when his men stood to their guns the frigate and the brig made off towards Livorno, which the British had captured six weeks before. Next day another British frigate arrived from Corsica, this time carrying a representative of the Provisional Government who had great difficulty in persuading the obstinate Dalesme that Napoleon had really fallen, that the exiled Emperor was on his way to Elba, and that in the interim the tricolour and the eagle must be replaced by the white flag and the fleur-de-lis. Although Dalesme grudgingly hauled down the red, white, and blue banner under which he had served so long, the news about Napoleon seemed so peculiar that he still feared a British trick to gain possession of the island; he had, moreover, been told to expect Napoleon on a French vessel, and when the *Undaunted* was sighted on 3 May he therefore kept his heavy cannon trained on the frigate until Ussher made a signal to parley and sent in Drouot and the other officers to explain the situation.

It was about six in the evening when the party reached Dalesme's office, where he hastily assembled the local dignitaries—Traditi, who was the mayor of Portoferraio, Balbiani, the sub-prefect who was responsible for the civil administration of the island, Dr Christian Lapi, who commanded two rather shaky companies of the National Guard, and a man called André Pons de l'Hérault, who managed the iron mines. It was a simple ceremony. Dalesme's doubts had been settled by the presence of General Drouot, whose integrity was a byword in the French army, and by the attendance of Campbell and Clam, and he accepted the orders from Paris which authorized him to surrender the island, together with all the guns and military stores. He then read aloud the letter of accession which Napoleon had drafted in Fréjus, and the little formalities which established a new kingdom were complete. The next step was an informal visit to the *Undaunted* to welcome its new king.

This was a strange occasion for the Elban officials who hurried out to the frigate, for they were being tested by rapid and bewildering changes of loyalty. Two weeks before they had been governing Elba on Napoleon's behalf, as part of the Empire; they had since passed through a half-hearted rebellion, a military mutiny, and a threat of a British annexation to find themselves agents of the Bourbon restoration; and they were now to serve Napoleon again, though in an altered and much-reduced state. 'It all seemed like a dream, a painful dream, a frightful dream', Pons de l'Hérault wrote afterwards. His feelings were particularly ambivalent. A plump, good-natured, versatile man of forty-two who had served at sea and then as a fellow-officer of Napoleon in the artillery before Toulon, he had always been a fierce republican and a critic of Napoleon's imperial pretensions. Yet he was an equally ardent patriot who worshipped Napoleon as General Bonaparte, the victor of Austerlitz and so many other triumphs. 'And here was I', he reflected as he went out to the *Undaunted*, 'going to see the hero who had given up his halo of glory!' He was by nature a man of enthusiasms and the excitement on this May evening was almost too much for him. 'My heart seemed fit to burst, my head reeled, and I trembled all over', he wrote, and his companions seemed similarly affected.

When they boarded the *Undaunted* they were given a perfunctory welcome by Bertrand, who seemed deeply depressed, and then Napoleon made a good entrance. 'He was carefully dressed', Pons noted, 'like a soldier ready for an official reception', with the chevalier's star of the Legion of Honour pinned to the green coat of a cavalry colonel which was always his favourite uniform. 'He was calm, his eyes were bright, he had a kindly look on his face, and he smiled in a dignified manner.' After General Dalesme had stammered out a few respectful words, with concurring mumbles from the rest of the deputation, Napoleon delivered a crisp address in which he spoke warmly of his love for France, reviewed the misfortunes which had brought him to Elba, and declared that he now proposed 'to consecrate himself to the welfare of the Elbans'. There was a short personal word for each man, and then the interview was over.

Until that morning Elba had been seething with discontent, as Campbell learned as soon as he landed. 'The spirit of the inhabitants is very inimical to the late Government of France, and personally to Napoleon', he wrote when he returned to the *Undaunted*; and as the man whom Castlereagh would hold responsible for Bonaparte's safety he was understandably worried about the risks that had to be run before the companies of the Guard reached the island. Yet he exaggerated the danger. The Elbans had learned to adjust to a sudden change of fortunes, and once they realized that Napoleon's arrival was bound to mean more money, more trade, notoriety, and excitement, their hostility could easily turn to enthusiasm. In any case, as Campbell wrote cynically to Castlereagh that night, in a place which is dominated by a garrison 'it is not difficult for the officers to produce cries of joy from the populace'.

The news that the great Napoleon had arrived spread quickly, and crowds lined the walls and quayside to stare at the British frigate which lay at anchor only a few hundred yards away. As Dalesme and the officials returned to an almost sleepless night in which they had to improvise everything for a royal reception next day, the black water was pricked by reflections from the candles placed by the citizens of Portoferraio in their windows as a gesture of welcome. A tiny kingdom was about to receive a great king.

6

THE KING OF ELBA

May, the local saying goes, is the month of Elba, when the hills are freshly green and the warm sun brings out the gleam of the broom and the sweet scents of the *macchia*; and on the first day of the month, in a traditional rite, the young men go from door to door to sing the sentimental serenade of the *Maggio* and to receive the sugary *corollo* cakes which symbolize goodwill. In 1814 the feast-day fell on a Sunday, and the celebrations were scarcely over before the Elbans had suddenly to prepare for an exceptional and even greater fiesta.

No one explained to them exactly why Napoleon had come to their island, or how he had acquired it as a private domain. As soon as General Dalesme returned from the *Undaunted* he drew up a statement which tactfully declared that it was 'the vicissitudes of human affairs' that had brought the Emperor to Elba, congratulated the islanders on their 'good fortune', and quoted Napoleon's own disingenuous comment, in the letter drafted at Fréjus, that he had himself chosen to live among the Elbans 'on account of their pleasant manners and the excellence of their climate'. This announcement, hurriedly printed in the night, was sent all over the island, and in some places it was pasted up to cover a recent proclamation in which Dalesme had denounced the rebellious Elbans as savages and threatened them with dire penalties. The printer Broglia also had to turn out a manifesto from Balbiani, equally vague and complimentary in welcoming Napoleon on behalf of the civil administration, and a flowery address from the Vicar-General, a bibulous Corsican named Giuseppe Filippo Arrighi who claimed some sort of cousinship to the Emperor. Arrighi likewise avoided the delicate matter of exile, referring only to the 'changing politics of Europe' and asserting that Divine Providence had given Elba this chance of 'glory and prosperity'. The prospect moved him to flights of hyperbole. 'Wealth will pour into the land', he told his flock in phrases calculated to appeal to the self-interest of a people which had so recently been in revolt against the exactions of the Napoleonic regime. 'Join in all the respectful sentiments of love for

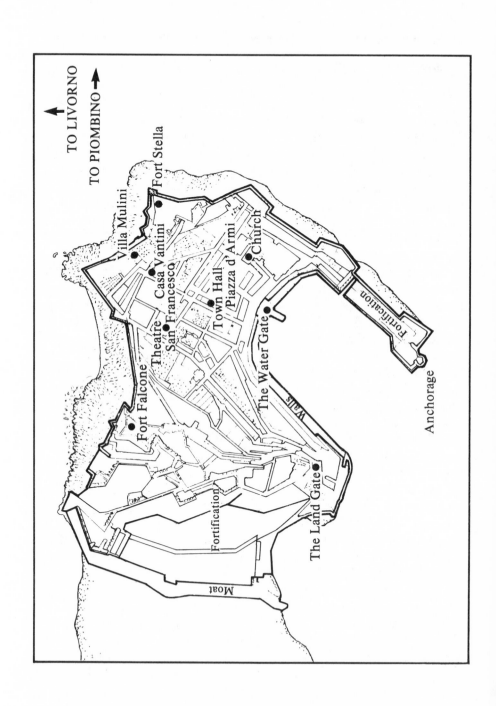

TO LIVORNO

TO PIOMBINO

Fort Stella

Villa Mulini

Casa Vantini

Church

San Francesco

Theatre

Town Hall

Piazza d'Armi

Fort Falcone

The Water Gate

Fortification

Anchorage

Walls

Fortification

The Land Gate

Moat

your Prince, who is more your father than your sovereign, and exult with holy joy in the bounty of the Lord, who has marked you out for all eternity by this auspicious event.'

Traditi, the mayor of Portoferraio, was too busy for such rhetoric for he had only a few hours in which to improvise a royal welcome. He persuaded his fellow-councillors to vote 3,000 francs for illuminations and a free issue of bread to the poor. He set men to work on a dais of painted plaster and gold paper which could carry a sofa as a temporary throne; he despatched more men to deck the streets with bunting and garlands of myrtle; and, because the little town lacked any ceremonial keys which could be presented to Napoleon, he took the keys from his own cellar and had them gilded for the occasion. He next had to find temporary lodgings for the new King of Elba and his staff. The best he could do for Napoleon himself was to clean out the unused upper floor of the town hall—commonly called the 'Biscotteria' because it had once housed a biscuit manufactoŕy—and to call on the more prosperous citizens to lend suitable pieces of furniture. Finally, he sent messengers about the island to ensure that all the notabilities were on hand next day and that the welcoming crowds were as large as possible.

There was one point of ceremony which had been on Napoleon's mind all the way to Elba. He needed a suitable flag for his little kingdom and the recognition of his shipping, and when he met the deputation he had asked for its advice. The eager Pons de l'Hérault thought up a design overnight and had a boatman deliver it to the *Undaunted*. Balbiani, who was less impetuous, merely provided Napoleon with a book which portrayed banners and coats of arms with Elban associations. From this source Napoleon took the white flag with a red diagonal which had been the standard of Cosimo de Medici when Elba was ruled from Florence; and he imposed three golden bees on the diagonal. It was variously suggested that he took the bees from an old print of Elba he had found in the library at Fontainebleau, that they represented peace, harmony, and industry, or that they stood for the three main townships of Elba, but in fact he had already included them in his arms as Emperor of France, believing that they were emblems used by the medieval kings. Napoleon himself said nothing to explain the heraldry except a sardonic remark that he had coloured the bees gold rather than blue in case the flag seemed to be a disguised tricolour, but whatever the reason for the design it served well enough, and the ship's tailor on the *Undaunted* hastily stitched up the first two flags—one to be run up over Fort Stella and the other to fly from the ship's barge as the new King of Elba went ashore.

★

Napoleon, who was up at dawn, spent a long time interrogating the harbour-master who had come on board to take the *Undaunted* in to anchor just off the mole; and while the boats began to unload his baggage he received some local dignitaries who wished to meet him before the official reception—Arrighi, eager to make something of his Corsican connection, Baccini, who hoped to be confirmed as president of the tribunal which served Elba as a court of law, and Colonel Vincent, an old acquaintance who had been commanding the army engineers in Tuscany at the time of the collapse and had been cut off in Elba.

Vincent stayed on board and at eight o'clock he went with Napoleon, Bertrand, Campbell, and Ussher on a trip across the bay to San Giovanni to inspect a large farmhouse Napoleon had noticed from the ship as a possible residence. As they were being rowed to the shore, Ussher recalled, Napoleon suddenly realized that he had no sword, and asked with something of a forced smile whether the Tuscan peasants were addicted to assassination. 'Evidently,' Campbell wrote in his journal, 'he is greatly afraid of falling in this way.'

By early afternoon the frigate was surrounded by small craft loaded with sightseers and serenaders. Ashore, the quays and walls of Portoferraio were packed with people waiting in the sun: men from the Elban National Guard and the French garrison were drawn up in two ranks in front of the Water Gate where Napoleon was to land; in the barge of the *Undaunted* fourteen men of the boat's crew sat in their best rig—white trousers, short blue jackets, red neckerchiefs, and square tarpaulin hats capping greased hair and pig-tails; and the yards of the frigate were manned by seamen to speed Napoleon on his way with the customary naval cheers. It was a modest occasion for a man who had entered so many great cities in triumph, but everything possible had been done to keep up appearances, and as Napoleon went over the side with Ussher, Campbell, Köller, Bertrand, and Clam, the cannon of the *Undaunted* began the punctuated firing of the royal salute. They were echoed by the crack of lighter guns on two French corvettes in the roadstead, by the boom of the shore batteries, the bells of the churches, and the shouts of the crowd.

As the captain's barge pulled smartly away from the frigate

Napoleon glanced at the flanking boats loaded with officers and members of his suite, as well as a party of British marines which, at his request, Ussher had provided as a temporary personal escort to ensure his safety. Then he stood impassively in the bow of the barge for the twenty minutes it took to make its way into the Darsena, the neat square harbour behind the mole, and towards the little square jetty that lay like an apron in front of the Water Gate that led into the town.

Napoleon had never been a physically imposing man and now his chestnut hair was thinning, his complexion was pasty, and his body, always too long for his legs, was thickening with middle age. Yet in his famous costume of green coat and white breeches, with a new Elban cockade for his hat, he set himself apart, giving the calculated impression that he was the Caesar of the age and that all were right to come and gaze at him.

That impression was too much for Traditi, who was trembling with stage-fright. As Napoleon landed the mayor came forward, bowed, presented his gilded keys on a silver plate, and realized that he could neither find the speech he had tucked in a pocket nor remember one word of it. Napoleon was not in the least embarrassed, and all through the afternoon he took the initiative and even prompted the nervous officials in their parts. He quickly touched the keys, murmured a polite phrase to Traditi, exchanged compliments with Balbiani, kissed a cross held out to him by Arrighi, and moved on to stand beneath a scarlet-and-tinsel baldachin of the kind used in religious processions which had been hurriedly furbished to give some dignity to the occasion. The rest of the party quickly fell in behind him. Bertrand and Drouot came next in line, followed by Campbell and Köller, Ussher and Jerzmanowski, members of Napoleon's staff, the officers from the *Undaunted*, General Dalesme from the garrison, civic notables from Portoferraio, and lesser lights from the other Elban communes. The procession had only to pass through the gate, which was the sole opening in the Grand Rampart, and then cross a small square, but the crowd pressed so closely upon Napoleon that the files of soldiers were barely able to keep order and Arrighi was reduced to pushing and shouting at those who blocked the way.

It was a far cry from Paris and the day in Notre-Dame when Napoleon had crowned himself in the presence of the Pope. The best that Elba could offer was a *Te Deum* in a parish church, decorated as if for a local saint's day and filled with simple soldiers and townspeople. It was also a far cry from the Tuileries, and the pomp of the Empire, to a reception in a makeshift throne-room in the town hall, where three violinists and two cellists entertained the guests as they jostled for an audience with Napoleon. But he took both these parochial events in regal style, behaving reverently in the church and greatly impressing the Elbans at the reception by his apparent knowledge of the island and his close interest in their problems. Like Ussher, who was astonished when Napoleon picked out an old soldier as a man he had decorated on the battlefield at Eylau, the Elbans were unaware of his knack of recalling quickly-learned facts or putting names to faces—a knack much helped by careful preparation and prompting by his staff.

Once the guests had left to join the celebrations in the town Napoleon looked through the apartment that Traditi had hastily furnished for him. It was inconvenient, it smelt of the gutters outside, and the singing in the square proved that it was noisy. 'The accommodation is mediocre, but I will arrange for something better in a week or two', he wrote tersely to Marie-Louise that evening, in a letter telling her of his arrival in Elba and complaining that he was still without news of her. 'That is my daily sorrow,' he added. 'Farewell, my dear; you are far away, but my thoughts are with my Louise.' Very few of these letters ever reached her, and Napoleon received only a handful of notes she sent him on her way to Vienna.

The light was failing as he rode out with some of his entourage a little way beyond San Giovanni, where the track to Porto Longone curved round the bay, looked for a while at his little capital silhouetted against the sunset, and cantered back into town.

★

'The Emperor appeared to be indefatigable', Pons de l'Hérault remarked, 'because he did only what he wanted, how he wanted, and when he wanted.' It is true that the task of ruling Europe had taught Napoleon how to conserve his energies, but in his first days on Elba he drove himself hard. 'I have never seen a man in any situation of life with so much personal activity and restless perseverance,' Colonel Campbell concluded after riding about the island with Napoleon. 'He appears to take so much pleasure in perpetual movement, and in seeing those who accompany him sink under fatigue, as has been the case on several occasions when I have accompanied him. I do not think it possible for him to sit down to study, on any pursuits of retirement, as proclaimed by him to be his intention, as long as his state of health permits corporeal exercise.' Curious by nature, and stimulated by novelty, Napoleon was so keen to explore his inheritance that he was genuinely busy from dawn until dusk, familiarizing himself with the problems of Elba and taking stock of its limited assets; and in any case an attempt to impose himself on his little kingdom was better than brooding on the malign twist of fortune which had brought him there.

His personal safety was his prime concern. On the first morning, in fact, Ussher was woken at 4.0 a.m. by the beating of drums and cries of 'Vive l'Empereur!' as Napoleon set off to look over the defences of Portoferraio. By the time he returned for breakfast at ten he had examined the town walls, peered into military storehouses, climbed the cobbled alleys which run up through Portoferraio to the forts like sets of staircases, walked through the barracks exchanging soldierly pleasantries with veterans from Marengo and Egypt and, with the eye of an artilleryman, sized up the strength and siting of the batteries. What he saw reassured him. On the whim of Alexander, the Allies had casually made the most crucial of all the decisions at the end of the war, installing him in a powerful fortress, with cliffs protecting it on the seaward side and an easily defended town and harbour under its guns. He was certainly safe from a raid by Barbary pirates—a threat that he feared might be secretly encouraged by the Bourbons or Metternich; and once his loyal Guard reached Elba and replaced the present garrison it would plainly be difficult to remove him from the island against his will. 'He has therefore made a better bargain than most people imagine', Ussher wrote to a fellow-officer on 20 May, 'and may be comfortable if his imagination will allow himself to be so, in any situation.'

On this tour of inspection Napoleon began to look for a permanent home above the noisome alleys of Portoferraio, where the air would be better and the privacy much greater. But General Dalesme's spartan quarters were too small for even a modest royal residence, there were no sizeable mansions among the tangle of fortifications, and when Napoleon proposed to convert the San Francesco monastery, recently a barracks, he was talked out of the idea by Bertrand. The Grand Marshal wished to have his own establishment when his pregnant wife Fanny arrived with the children, and he feared that Napoleon might expect them to share this makeshift palace. The next choice was a dilapidated building on the ridge between the two forts. Simply called I Mulini, after two windmills which had stood beside it until they were demolished in 1808, it had been built almost a century before as a four-room cottage for the gardener of the governor, with outhouses which had served as a granary and stores for the garrison; and it was currently used as a mess by the officers of the engineers and the artillery. Although it had little to commend it except an ideal position, Napoleon saw its possibilities. Within two weeks, he decided, a part of it could be made habitable, and by the early autumn it could be turned into an agreeable and reasonably substantial villa, with a garden on the cliff-top overlooking a secluded beach, and a view across the sea-passages between Tuscany, Corsica, and France.

Napoleon was an impulsive man, eager to act upon any attractive idea as soon as it occurred to him, and his mind ran easily from plans for a modest palace in Portoferraio to the notion of a rural retreat. In the early afternoon, apparently unwearied by his six-hour inspection before breakfast, he took Ussher on a long ride into the countryside to look for a property that might suit this purpose, as in his heyday St Cloud had served as an alternative to the Tuileries. Even then he was not content to relax, for that evening he summoned Pons de l'Hérault—who preferred to live with his family in Portoferraio than in the rougher context of the mining community at Rio—and despatched him on a ten-mile ride through the darkness to prepare for a royal visit to the iron mines on the following morning.

The visit to Rio was the first chance of an outing for the group which had borne the strain of the journey from Paris, the uncomfortable crossing in the *Undaunted*, and the establishment of Napoleon on an island whose inhabitants had so recently revolted against him. Bertrand and Drouot were always at his side, and Köller, Clam, Campbell and Ussher, were also in the party which left at 6.0 a.m. and crossed the bay in Ussher's barge to avoid the long ride round the head of the roadstead. Landing at the tiny village of Magazzini, where sturdy Elban ponies were waiting to carry them eastward over the shoulder of the mountain to Rio nell Elba and Rio Marina, they were almost upon the villages before Pons was ready for them.

In the succeeding weeks, as Bertrand imposed a gimcrack copy of the imperial style upon the island, Napoleon began to go about in a grander cavalcade, with orderlies, gendarmes, chamberlains, and local notables flanking the personal staff who made up his little court. For Bertrand was a stickler for detail in matters which affected his master's dignity, issuing precise instructions about Napoleon's reception in each commune, including the construction of triumphal arches and the deployment of people who knew what to shout when they saw their sovereign. Despite the short notice, Bertrand had given useful advice to the nervous Pons, who declared that his lifelong attachment to liberty, fraternity, and equality had left him a complete novice in such ceremonial matters.

He had done his best to make a good impression as the royal party came to the two villages of Rio, set in a wild landscape in which the red spoil heaps of the iron workings contrasted vividly with the blue water of the Piombino channel. There were arches of chestnut boughs across the track leading down to Rio, a large and hastily sewn Elban flag, a bevy of pretty girls to present the visitors with flowers, lines of miners, pick on shoulder, to form a kind of honour guard, and a ragged salute from the culverins on the small ships that were used to transport the ore to the mainland. Even the fishermen from Rio Marina, who had been sent out to make a catch for the official reception, had done well and netted an unusually big *dentice*.

Yet one small thing after another went wrong. As Napoleon rode into the village he realized that the miners were cheering the popular Pons rather than himself—the cries of 'Viva il nostro Babbo!' were noticeably louder than the rehearsed shouts of 'Viva l'Imperatore!'—and he turned to Pons with a bleak smile to say 'So it's you who are the real prince here'. At the *Te Deum* in the village church, Ussher noted, 'the officiating priest did not seem to understand his business', and when Pons led Napoleon to his own house for breakfast he discovered that the man who had been told to deck it with flowers had chosen fleurs-de-lis. This display of the Bourbon emblem greatly distressed Pons who began to splutter and muddle up his forms of address, calling Napoleon 'Monsieur le duc', 'Monsieur le Comte' and even plain 'Monsieur', and it irritated Napoleon, who pointedly ignored his host and made it an uncomfortable occasion for everyone in the party. Years of soldiering and politics had made Napoleon an easy mixer by comparison with other sovereigns, but he was a sensitive man on small points of dignity. He only relaxed when breakfast was over, and Dalesme had formally assured him that Pons remained a republican and had intended no slight by the show of lilies.

All the same there was still tension between them, as Pons

parried Napoleon's hints that he might remain as manager of the mines, for the differences between them were actually due to something more than the gaucheries of the reception or the tussle between two proud personalities. What was really at stake was the future of the iron mines and the substantial revenues they produced.

In 1809, when Pons was sent to Elba to revive the mining operations at Rio, Napoleon himself had given their profits to provide pensions for soldiers he decorated with the Legion of Honour. Now times had changed and he needed the money for his own purposes, he faced the embarrassing task of taking it away from his old comrades-in-arms. Pons was crucial to this operation, for he was both the competent manager of the mines and a conscientiously honest person who felt that he was responsible for transmitting the income to the pension fund. Of course Napoleon could easily have seized the cash in hand, which had accumulated to about 500,000 francs during the months when the British blockade cut communications with France, and diverted the new revenues into his own Treasury without further argument, on the grounds that the Treaty of Fontainebleau had made the mines crown property. Yet such drastic action to assert his claim to the most profitable enterprise in Elba would have been unwise. He always preferred to observe the forms of law, wherever that was possible without compromising his intentions, and in this case there was an additional reason for caution. An open quarrel with Pons, whose control of the mines made him one of the most influential men in Elba, would have been a very bad start to his reign; it would have upset the Elbans and it would also have driven Pons away at a time when he sorely needed experienced men to help him run the island. He had gone to Rio so soon after his arrival, in fact, to woo Pons by the alternation of hauteur and patronizing charm which had won over much more recalcitrant opponents than the good-natured and naive mine manager, and the technique of conversion was so effective that years later, when Pons wrote his valuable but flattering account of Napoleon on Elba, he still felt a need to describe it with detailed ingenuous candour. Pons, quite simply, was spellbound by the presence of the Emperor and within a few days, after Drouot had called with a formal offer of continued employment, he entered Napoleon's service and acquiesced in the protracted but face-saving negotiations which eventually gave Napoleon the money he wanted.

From that point of view it was a very promising day, and it put Napoleon in good spirits. As the party made its way back he had to

General Drouot

General Cambronne

General Bertrand

Pons de l'Hérault

This portrait of Colonel Sir Neil Campbell is
apparently the only surviving likeness taken
from life

Count Neipperg

Marie-Louise and the King of Rome, an
engraving of a painting by Isabey. Napoleon
took a miniature of the painting to Elba

be dissuaded from climbing up to the ruined castle at Volterraio, with its spectacular view across the bay to the range of mountains rising beyond Monte Cappane, and Campbell noted with irritation that, at his insistence, Ussher's barge made detours to 'the watering-place, the height opposite the citadel on which he proposes to establish a sea-battery, and a rock at the mouth of the harbour on which he also thinks of placing a tower'. For the first time Napoleon seemed to feel confident about his prospects in his small island kingdom.

★

At dinner that evening, indeed, Napoleon was quite excited about the future. 'Already he has all his plans in agitation', Campbell noted, 'such as to convey water from the mountains to the city, to prepare a country house, a house in Portoferraio for himself, another for Princess Pauline, a stable for 150 horses, a lazaretto for vessels to perform quarantine, a depot for the salt, and another for the nets belonging to the fishery of the tunny.' It was always Napoleon's habit, when he came to a fresh place, to let his fancy run on plans for improving it, regardless of practical need or expense, and because he had great power many such things were set in train as he travelled about Europe. But since the schemes he devised during his first weeks on Elba would have been sufficient to absorb his remaining resources for a lifetime most of them were soon abandoned. It was too difficult to imitate the Romans and bring water to the town by aqueduct or (another passing notion) to pump seawater to the top of the alleys in Portoferraio to provide cleansing streams down the gutters. It was too expensive to build proper roads for his carriages. It was to prove impossible to colonize the island of Pianosa. All the same he pressed some of his ideas against the advice of Bertrand, especially where they served his own comfort or interests. To make a stable 'merely for the convenience of his horses', Campbell noted in a dour phrase, he commandeered the warehouse of Pellegrino Senno, the wealthiest man on Elba, who had made his fortune from the tunny fishing. This action, Campbell added, was 'likely to cause some disgust', and he thought it demonstrated 'how little Napoleon permits reflection to check his desires'.

On the day after the visit to Rio, moreover, Napoleon delivered his first comprehensive order to General Drouot, who had been persuaded to act as governor of the island. 'Hoist the new flag in every village and make the occasion a kind of holiday,' Napoleon wrote in a set of instructions born of long experience of military occupations and changes of government. 'Call a meeting of the naval commissioner, the harbour-master, the commander of the French ships in the harbour ... arrange for me to see anyone who can brief me about the civil administration, customs, taxes, sanitary matters and the management of the port ... find a reliable man who can take charge of stores and supplies ... find out how many of the French garrison wish to remain ... make sure that the gendarmerie is efficiently organized—that's very important ... see the sub-prefect about a store of grain that he says belongs to the Grand Duchess of Tuscany and take it over as crown property ... disband the coastal artillery companies, they cost too much, and anyway they are useless ...' This order, like so many dictated by Napoleon, was rambling in structure, conversational in tone, and packed with petty matters which could well have been left to the painstaking Drouot or other subordinates. For a man of such large designs Napoleon had always been a fidget about detail, a worrier; and from the moment he arrived on Elba he began to occupy himself with day-to-day business less suited to a king than a clerk.

There was some formal business on Sunday 8 May when the British frigate *Curaçoa* came in from Genoa, carrying Mr Locker, the secretary to Admiral Sir Edward Pellew. For it was Locker's duty formally to present Napoleon with a copy of the preliminary peace treaty between the Allies and the Provisional Government which had been signed in Paris on 23 April. Napoleon read the document in the presence of his aides, Köller, Campbell, Ussher, and Captain Towers of the *Curaçoa,* and then returned it to Locker with no comment beyond an expression of thanks, as if to imply that he was now indifferent to the affairs of France. When Köller sailed next day on the *Curaçoa* on the first stage of his journey back to Paris, he took with him a letter to Marie-Louise which gave the superficial impression that Napoleon was resigned to his new conditions of life. 'I have been here five days,' he told his wife with full awareness that the Emperor of Austria and Metternich would see whatever he wrote, 'I am having prepared a nice house with a garden where the air is good, and I shall go there in three days. My health is perfect. The inhabitants seem good people and the country is agreeable. What I want is to have news of you, and to know that you are well, for I have heard nothing since your courier reached me at Fréjus. Goodbye, my dear. Give a kiss to my son.'

Campbell was not much impressed by this show of settled intentions. He was, in particular, uneasy about Napoleon's efforts to secure furniture from Italy for his new house, which he thought might be a cover for secret communications to his family and his adherents; and although he gave a passport to one of the royal valets who was sent to Milan to negotiate for some chattels belonging to Pauline's husband, Prince Borghese, at the same time he asked Köller to take a despatch to Lord William Bentinck to ensure that the man was watched after he landed in Genoa.

With Köller's departure Campbell was the only Allied representative left on Elba, and he was in a curious position. Napoleon was apparently well-disposed towards him, spoke freely in his presence, and was insistent on his company. On the evening that Köller left, for instance, Napoleon talked to Campbell and Ussher for over three hours about the last stages of the campaign in France, and he went on to recall his plans for outbuilding the British fleet, outmanoeuvring Nelson, invading the southern counties and marching on London. 'His thoughts seem to dwell perpetually upon the operations of war', Campbell remarked on 26 May, after listening to several discourses of this kind. They all pointed, Campbell said, to the worrying conclusion that 'if opportunities for warfare upon a great scale and for important objects do not present themselves, he is likely to avail himself of any others, in order to indulge this passion from mere recklessness'.

In all these martial recollections, as in the long conversations on the journey across France, on the *Undaunted*, and during the gadabouts on Elba, Campbell repeatedly sensed Napoleon's hope for an eventual change of fortune which sustained his soldierly distrust of the Corsican Ogre. Sometimes, he noted in his journal on 27 May, Napoleon 'throws off all restraint, and expresses himself so openly as to leave no doubt of his expecting that circumstances may yet call him to the throne of France'; and he quoted several phrases to support that opinion—'France is humiliated . . . the minds of the people are very unsettled . . . the greater number of the population and the whole of the army are for himself . . . the family of the Bourbons will be driven out in six months . . .' Napoleon, he decided, 'certainly regrets that he gave up the contest'.

Before Napoleon had been on Elba a week Campbell was convinced that a man of such energetic temperament, with such memories of glory, and such a talent for dissembling in politics and war, was bound to find his little kingdom too small a stage for his talents. It was this troubling conviction, indeed, that persuaded Campbell that he must go beyond Castlereagh's explicit orders, which merely authorized him to protect Bonaparte from 'insult and attack', and that he must start playing a different role. Napoleon, he told Castlereagh in a confidential despatch which Köller carried back to Paris, had repeatedly invited him to stay on at Portoferraio, at least until the English transports ferried Cambronne and the Guard across from Italy. He had now accepted this useful invitation, since it would give him an opportunity 'to obtain information of the intercourse which Napoleon or his agents hold on the opposite coast of Corsica'.

Campbell thus put himself in an impossible situation, for he had neither the personality nor the training for such clandestine activities, and his government casually left him there without any clear responsibility or support. He was simply an anxious amateur, pitting himself against a man who was a master of deception, and creating the illusion that Napoleon was being carefully watched even if he was not being carefully guarded. It was an understandable decision but it was to have fateful consequences.

7

A VERY SMALL ISLAND

A week after Napoleon landed in Elba he rode up Monte Orello, a hill in the centre of the island, and there surveyed over his whole domain. 'Well,' he said ruefully to Campbell as he looked round the circling cliffs, promontories, and beaches, 'it certainly is a very small island.'

He soon saw it all, for on 18 May he set out on the last of the outings which served as a combination of a royal progress and a military reconnaissance. Accompanied by a jingling entourage of gendarmes, Elban dignitaries, British officers, and his own staff, he first went westwards, trotting through triumphal arches into Marciana Marina (where he had been burned in effigy less than a month before) to celebrate a *Te Deum* in the church. Next day he climbed the shoulder of Monte Capanne to reach Marciana Alto and Poggio, and then crossed the spine of the island to the little port of Campo lazing on the curve of a south-facing sandy beach. And on the third day he sailed his party across from Campo to picnic on Pianosa, the small flat island fifteen miles to the south-west which the Allies had not seen fit to include in the territory of Elba.

Although it had once been a flourishing Roman colony, with a palace, temples, a theatre, and baths, Pianosa was now uninhabited. Time and again the Elbans who tried to settle it had been butchered or carried off by corsairs, and since 1806, when a French garrison had been forced to surrender to a British frigate, the Elbans had used it simply as a pasture for running wild horses. For Napoleon, any extension of his tiny kingdom was welcome, and after he had ridden over its few square miles he had devised a scheme for its future. He would send troops to turn it into a defensive outpost, peasants to make use of its fertile soil, and fishermen to take good catches from its dark clear waters. 'This occupation will make streams of ink flow in the chancelleries of Europe,' he jokingly remarked as he gave Drouot the necessary orders. 'They will say that I have not renounced my habit of conquest!'

As soon as Napoleon returned to Portoferraio he moved into the Villa Mulini, where the workmen had been hurried to make part of the building habitable while the rest was being altered to provide him with the most modest of palaces. The house itself was pleasant, though far from grand, and even when the builders had added a floor to the central block and carried out other improvements— acting under Napoleon's characteristic instructions that the work should be done 'at the least possible expense'—the property was more appropriate for a garrison commander or a rising merchant than a fallen emperor. By the time it was completed at the end of August it was no more than a simple two-storeyed structure in pinkish-white plaster, with green-painted shutters, which ran for 110 feet along one side of a space that had been cleared for car- riages to turn; and before carriages could actually be brought up to the door Napoleon was obliged to use troops and miners to dig a long tunnel through the rock. Behind the house, on the seaward side, a wing ran back towards the parapet at the cliff edge which contained a treeless garden with gravelled paths. There were in all some thirty rooms, most of them small and some tiny, and there was also a barrack of lodgings just below the house which accommodated guards and servants.

Napoleon himself lived on the ground floor in the central part of the Villa Mulini, where he had a bathroom, bedroom, a study, a library, a fair-sized salon with french windows opening into the gar- den, and ante-rooms for his personal staff and the orderly officers. At the same level, in an adjoining wing, were his kitchens, and a large room set at an angle to the house in which he installed a mov- able partition so that it could be used for dinners or musical and theatrical entertainments as occasion required. Upstairs were the apartments he prepared in the hope that Marie-Louise and the King of Rome would join him, and these flanked the only sizeable room in the building—the Grand Salon, about forty feet long and stretching across the full thirty-foot depth of the house, so that its windows opened on one side to the sea and on the other to the har- bour and the mountains. It was an attractive, bright, and decorated chamber which contrasted with the cramped and unimpressive suite Napoleon occupied below. For weeks after his arrival at Por- toferraio he continued to sleep on the collapsible bed which he had carried from one end of Europe to the other, and when he set up a summer retreat at Marciana Alto or spent a few days at Porto Lon- gone he did no more than camp there with the minimum of domes- tic comfort.

Furnishings for the Villa Mulini were indeed a problem, for the island was a spartan place with hardly any well-appointed houses or even the cabinet-makers and upholsterers who could supply the deficiency. Napoleon immediately looked across to Tuscany, until recently ruled by his sister Elisa Bacciochi and her husband, and decided to strip their vacant Piombino palace of furniture, hangings, and anything else that might be useful. A few weeks later, as Napoleon sardonically remarked, he had another chance 'to keep things in the family', for his sister Pauline's estranged husband Prince Borghese had been obliged to leave Turin for Rome and the ship carrying his furniture was driven into Porto Longone by a storm. This shipment, too, was requisitioned, with a meticulous but unpaid inventory, and Napoleon rounded off his acquisitions by spending a few thousand francs on chattels taken over from officers of the departing French garrison. Altogether he had collected a set of oddments which ranged from the ornate canopied and gilded bed, ringed by carved swans, on which his sister Elisa had so notoriously entertained her lovers, to the faded drapes and threadbare carpet which Fleury de Chaboulon noticed when he visited the house early in 1815. 'I remembered the luxury of the imperial palaces', he wrote, 'and the comparison drew a deep sigh from me.'

Napoleon had a similar problem with his personal possessions. He had managed to bring some table silver with him, but almost all the valuable tableware and ornaments had been sent off in convoy to Blois when Marie-Louise fled from Paris; and he had lost his dress uniforms, state robes, and personal linen in the same way. Both he and his men were so short of clothing in fact, that in June he took the opportunity to buy the cargo of an English ship that had been captured before the war ended. There was cotton for pants and vests, green cloth for liveries, blue serge for his soldiers, and even percale for his sheets. His small but interesting library was also assembled piecemeal. He had picked out a good many books at Fontainebleau, he had even bought some as he passed through Lyons and Fréjus and others were ordered from Rome, Florence, Livorno, Venice, and France—standard editions of classical authors, collections of La Fontaine, Montaigne, Rousseau, and Voltaire, scientific, historical, and military texts, a forty-volume set of the fairy stories which he had always enjoyed reading, and a number of volumes which had been penalized by his imperial censors in the past and yet seemed surprisingly innocuous when he came to skim through them himself.

★

Exiles must make do with very different standards than emperors, and the unpretentious Villa Mulini set the scene for a royal household which was sadly diminished in both quality and size. It was, in fact, a court more suited to the stage of an opera-house than the corridors of a real palace; for only a few experienced principals were needed to govern an island as small as Elba, and many of the lesser parts, thought necessary for appearances, could be filled by local amateurs. Campbell remarked on 'the ridiculous nature' of a reception where the guests included the seamstress who had just repaired his uniform jacket.

There were some indispensable men in the party which had come from Fontainebleau. Bertrand, an able and conscientious administrator, could manage all Napoleon's personal affairs and keep control of the civil government as well. Drouot had seen so much service that he could take responsibility for the island's defences and still keep part of the day free for his studies in philosophy and theology; and he could count on unwavering assistance from Cambronne, due to arrive with the Guard and take command of the garrison in Portoferraio. Guillaume Peyrusse, the cool young man who had gone off to Orléans and Rambouillet and brought back enough money from Marie-Louise to keep Napoleon temporarily in funds, was an equally trustworthy treasurer. Rothery, one of Napoleon's secretaries, was also in the group to come over in the *Undaunted*; so were Callin the Comptroller of the Household, Sostain the Master of Ceremonies, the Chief Cook, the Chief Baker, and a score of valets, footmen, and other lackeys. But there were deficiencies even in this travelling household, since so many of his regular attendants had abandoned him in defeat. Captain Deschamps, who had come over as one of the grooms of the bedchamber, was simply a rough-mannered though affable man who had begun his career as a gendarme, and another was Captain Baillou, an old soldier better fitted for life as a campaigner than as a courtier. Foureau de Beauregard, who became the butt of everyone's humour at the Villa Mulini, was a pompous and gossipy veterinary surgeon from the imperial stables who had been promoted to court physician when the eminent Corvisart had considered it more agreeable to accompany Marie-Louise to Vienna than to go with Napoleon to Elba. His assistant, the apothecary Gatti, was an amiable but unqualified young man who had attached himself to Napoleon's entourage just before the journey across France.

The Elbans who filled the remaining posts were a mixed lot, too.

Though Napoleon was eager to conciliate the islanders by appointing local men to public and honorific positions it was hard to find suitable candidates in such a small and impoverished place. Three of the four court chamberlains, it is true, came from the small group of families with some claim to gentility. Dr Christian Lapi, a man with more charm than talent, and the part-time commander of the National Guard companies on Elba, was given the additional task of supervising the roads, bridges, and forests. Vincenzo Vantini, who served as the public prosecutor, was a witty and indiscreet person with claims to an aristocratic lineage which contrasted with his fervent support of republican principles; according to Pons he had 'run through his family fortune by indulging in excesses that were inexcusable'. Mayor Traditi of Portoferraio was an upright spokesman for the town's merchants and traders. But the fourth chamberlain was Gualandi, the self-important mayor of Rio. 'He was nothing,' Pons said, 'less than nothing, and should never have crossed the Emperor's door.'

These four chamberlains were also appointed to the Council of State, a body with a high-sounding title and few functions which Napoleon set up to give a semblance of constitutional procedures to his regime. In addition to Bertrand and Drouot, who effectively governed Elba between them, it included Pons; Giuseppe Balbiani, the Tuscan with a large family who was happy to continue as an administrator whether he was called a Sub-Prefect of France or the Intendant of Elba; Arrighi, the Vicar-General; Bartolini, the Arch-Priest of Capoliveri; Baccini, the Genoan lawyer who served as the President of the Tribunal; a couple of men who lent money and called themselves financiers; and a small landowner from Marciana called Sardi.

Napoleon had been hard put to find a dozen reasonably competent men to sit on his Council of State, and once they had been appointed it was even harder to fill the remaining posts. He was lucky to meet an old Corsican acquaintance named Poggi di Talavo on the island, for Poggi had served for a time as a judge, and apart from his qualities of loyalty and intelligence he had the gift of making people confide in him which made him into an effective head of the secret police and Napoleon's chief source of information. Poggi often stayed late at the Villa Mulini chatting to Napoleon, who loved to hear gossip about his associates and had long ago fallen into the habit of using informers to increase his power over them. But he had to entrust the command of his small fleet to Lieutenant

Taillade, a vain and pedantic fair-weather sailor who had left the French navy and become captain of an Elban schooner after he had married into one of the better local families. Captain Paoli, another of the Corsicans who stayed in Elba in the hope that they could better themselves by serving their famous fellow-islander, and who was put in charge of the gendarmerie, was so cravenly sycophantic that whenever Napoleon asked him the time he would reply 'Whatever time Your Majesty pleases!', and Pons remarked that Carlo Perez, one of the five young Elbans recruited to serve as orderly officers under the blustering and quarrelsome Major Roule, was so foolish that he was 'nearly an imbecile'. Too many of those appointed were almost comically incompetent, and their engagement made it clear to an outsider such as Colonel Campbell that the Elban regime was little more than a façade, erected to give Napoleon a face-saving pretence at royal dignity and, hopefully, to persuade both the Elbans and the outside world that he had settled down to live like any other Italian princeling.

To some extent that effort succeeded. 'The principal inhabitants of the island', Campbell wrote on 15 May, 'have also been impressed with the opinion that his possession of the island will afford them extraordinary resources and advantages; which opinion has extended itself to all classes.'

A host of servants was deployed to the same end. When local staff were added to the men who had come with Napoleon from France there were more than sixty people at work in and around the Villa Mulini. There were thirteen in the crowded kitchens, and nineteen more about the house; there was a director of music from Florence, who disposed of a French pianist and two singers; a head gardener of some ability called Claude Hollard, who had previously worked for Elisa Bacciochi in Piombino and now supervised two other men; there was an Elban mistress of the wardrobe, Signora Squarci, who was the wife of one of the assistant mayors in Portoferraio; and there were another twenty-two men working in the royal stables, quickly converted from a storehouse to accommodate Napoleon's favourite horses and the extraordinary array of wheeled vehicles which the Guard had brought on the long march from Fontainebleau.

Napoleon's household was short of so many essentials, so ill-balanced in people and resources, that these well-stocked stables seemed all the more incongruous—especially in an island where there was scarcely a passable road, and the only town of any size was built on a grid of steep cobbled passages. In Elba one went by horse and mule, or walked, and even Napoleon seldom drove more than two or three miles into the country. Yet by the end of the summer, after his mother and sister had arrived, he had imported a total of six *berlines*, two open carriages, a landau, two post-chaises, a yellow cabriolet that would have looked smart in Paris, two hunting-carts and a number of baggage-waggons. To keep all this transport and his horses in prime condition he employed a chief saddler, a harness-maker, a veterinarian, a farrier, a head groom and two assistants, a coachman, eight postillions, ten stablemen, two troopers, two locksmiths, two carpenters, and a tailor for the liveries of the men: there were gold-braided green coats and red capes for the coachmen and purple jackets and black caps for the outriders, brass imperial eagles and the Elban bees on the harnesses. All in all the domestic establishment would have been the most imposing aspect of Napoleon's little court if it could have been displayed to good advantage in a town palace or a country estate, but the Villa Mulini, huddled up against Fort Stella, was too small and inconvenient for such pomp. The contrast merely drew attention to the pathos of this attempt to maintain the trappings of power when the substance had been lost.

*

Napoleon had a reputation as a hard worker, capable of combining restless physical activity and mental effort; there were many anecdotes about his ability to dictate to two or three secretaries at once, to talk late into the night and to rise at dawn. The efforts he made to establish himself during his first weeks on Elba were certainly a good illustration of the way he could turn his attention and his energies to a number of different tasks at the same time. It is clear that the closer he got to the working habits which had been necessary to rule an empire and command a great army, the more cheerful he became, and as the hot summer wore on he seemed quite absorbed in the business of making Elba fit his fancy. 'He was like a thorough-bred gamester', Sir Walter Scott said a few years later, 'who, deprived of the means of depositing large stakes, will rather play at small game than leave the table.'

The stream of orders continued, as if Napoleon really believed that he could convert his poor sun-scorched island into a model kingdom. Lapi was given instructions to care for the forests, and to survey land on which wheat might be grown, for Elban farmers only raised a small part of the needed supply. An engineer was to be recruited at a salary of 1,800 francs to supervise work on the roads and improve the neglected salt marshes. Trees, especially the useful mulberry, which Napoleon's father had once planted to good effect in Corsica, were to be imported during the autumn months and planted by the single road from Portoferraio to Longone, and there were precise orders about the water supply and sanitary system. Napoleon asked for a weekly report on the level of the rainwater cisterns, one of which was to be reserved for the Villa Mulini and his garden; horses, he insisted, were to be watered only at one of the wells in the Ponticello barracks; a pump was to be set up at another of these wells to provide water to wash the alleys of Portoferraio; street-cleaners were to be appointed to patrol the town and prevent the citizens throwing garbage and ordure directly out of their doors; all householders in the capital were ordered, on pain of a punitive tax, to provide a latrine which could be emptied every night, and a drain was to be run into the sea to carry the waste from the San Francesco barracks. Napoleon had always been a stickler for cleanliness, and the few days he had spent in the odoriferous centre of Portoferraio had shown him the need for drastic measures if the town was to be made reasonably pleasant and healthy. 'The island of Elba had suffered much and long', Pons said with characteristic hyperbole in his memoirs, 'and the picture of prosperity which emerged after the Emperor's arrival at Portoferraio was little short of miraculous; the list of works, some in preparation and some in vigorous execution, but all directed to the greater Dignity of Man, displayed in a radiant light the supreme creative power of genius.'

But such reforms, if they were ever carried out, would be useful enough to the Elbans, and they were immediately useful to Napoleon. Like the details of rebuilding and staffing the Villa Mulini, or finding several hundred palliasses for his soldiers, or calculating whether ten sous would be a sufficient daily allowance to provide men in the military hospital with eggs and other nourishing food, they were grist to his active mind; 'his amusement', Scott said, 'consisted in constant change and alteration'. He appeared to need a multitude of small problems to keep himself alert, or perhaps to fend off anxiety at a time when he was concerned about more important matters but was unable immediately to do anything about them. During his first weeks on Elba, for instance, he busied himself with all sorts of trivialities while he worried about the problem of defending the island once the French garrison was repatriated and Ussher had taken the *Undaunted* back to normal naval duties; though he soon began to make plans for a small army and a tiny fleet he knew that everything depended upon the arrival of the Guard. Even with his seasoned and devoted veterans it would be difficult to organize an effective defence of the island, and without them it would be impossible. If they were delayed, or if the Allies changed their mind and found some reason to prevent them crossing to Elba, he would have no security at all.

Campbell tried to reassure him by insisting that the Allies would respect the clause in the Treaty of Fontainebleau which allowed a battalion of the Guard to join him, by arranging for British ships to ferry the men from Savona to Portoferraio, and by trying to time their arrival to coincide with the French garrison's departure. But on the morning of 25 May, when Napoleon knew only that the Guard had at last reached the Italian coast, Captain de Montcabrie— whose escort services Napoleon had rejected at Fréjus—came into Portoferraio from Toulon in the frigate *Dryade*, with orders to ship Dalesme's units back to France; and he brought with him the brig *Inconstant*, which was to be turned over to Napoleon in place of the larger corvette which had been specified by the treaty. Montcabrie's arrival, and the evasion of treaty obligations implied by the substitution of the smaller vessel, intensified Napoleon's anxiety. He now feared that he would be left without troops to man the forts and with only the brig to protect the approaches to the island.

That afternoon, however, Captain Ussher came in with better news. At Napoleon's request he had taken the *Undaunted* back to Fréjus in the hope of collecting his sister Pauline and bringing her to Elba, and though both he and Captain Towers in the *Curaçoa* had failed to find her, for she had already left in a Neapolitan frigate sent by Murat, he reported that the British transports carrying the Guard had been held up by a severe storm and that they were on their way. Napoleon stayed in the garden of the Mulini until it was dark, looking northward in the hope of sighting the flotilla, and later that night, after Ussher had dined with him, they went back to keep watch at the parapet. Near midnight, with the aid of Ussher's night-glass, they finally glimpsed the incoming ships, and Napoleon took a few hours rest until he went down at dawn to see them come into the roadstead.

The Guard had made a remarkable march of nearly a thousand miles in a month, proudly flaunting the tricolour through towns decked with Bourbon flags and occupied by Austrian regiments, reminding Frenchmen of past victories and their enemies of old defeats. They had been forced to abandon their cannon on the snow-bound pass over the Mont Cenis, but they had cossetted Napoleon's chargers and manhandled his baggage-waggons and carriages over the mountain tracks, and now they had come down out of the Alps into Italy as if they carried the last vestiges of French glory on their eagles.

Early next morning, Ussher recalled, Napoleon was waiting impatiently on the quayside long before the ships were worked into the harbour, and to pass the time he went on board the *Dryade*. This was a curious and impulsive breach of etiquette (since Captain de Montcabrie had not invited him and the frigate flew the Bourbon colours) which Campbell concluded was 'certainly done to try the disposition of the French navy, and to keep up a recollection of him in France'; and the French seamen in fact greeted him with spontaneous cries of 'Vive l'Empereur!' which were echoed by the *grognards* on the nearby transports who were waiting to go ashore. At eight o'clock they began to file on to the quays, where Dalesme and a guard of honour from the garrison were waiting to greet them, and as Cambronne came up Napoleon embraced him. 'Cambronne,' he exclaimed with evident relief, 'I have passed many bad hours waiting for you, but at last we are again united, and they are forgotten.' It was a moving scene, as this remnant of the Grand Army marched off with colours flying and drums beating to the Piazza d'Armi, and many of the veterans wept as Napoleon addressed them. 'Live in harmony with the Elbans,' he said, grandly overlooking their recent rebellion. 'They also have French hearts.'

The infantry moved off to occupy the San Francesco monastery, where the old campaigners made jokes about the new life in which Napoleon apparently expected them to live like monks. The grenadiers were sent up to Fort Stella, where Cambronne as garrison commander moved into a room next to Drouot's apartment, with a window looking down into the garden of the Villa Mulini sixty feet below, and the troop of Polish lancers who had remained passionately loyal to Napoleon were despatched to the height of Fort Falcone, where they were soon to be turned into artillerymen, there being little use for cavalry on Elba beyond ceremonial escort duties.

As soon as the ceremony was over Napoleon went back to the harbour, where he spent the whole day in the hot sun talking to

Cambronne and watching while the crew of the *Undaunted* helped
to unload the stores, horses, and baggage from the transports. The
work was completed by late afternoon and Napoleon then went on
board to make a farewell speech of thanks to the crew of the frigate, which was due to sail a few days later. 'Those fellows', he
said to Ussher, pointing to some Italian sailors, 'would have been
eight days doing what your men have done in so many hours; besides they would have broken my horses' legs, not one of which has
received a scratch.'

The reunion with the Guard had put Napoleon in an excellent
mood. He sent the crew of the *Undaunted* a large present of wine
and money, gave Ussher a snuff-box with his portrait set in diamonds, and was so gracious that even the cautious Campbell was
impressed. 'Napoleon speaks most gratefully to everyone of the
facilities which have been granted to him by the British Government,' Campbell noted smugly in his journal that night; 'and to
myself personally he constantly expresses the sense he entertains of
the superior qualities which the British nation possesses over every
other.'

*

Napoleon's remark, set in the context of so many arrivals
and departures, set Campbell thinking about his own position.
After all, he remarked, he had only outstayed the other commissioners at Napoleon's own request and to procure the naval facilities for
which he had just been thanked and he considered that he should
either leave with Ussher when the *Undaunted* sailed, or make a de-
cent offer to do so, even though the interests of Britain and the
other Allies might be better served if he stayed close to Napoleon
'in order to judge of his intentions'.

Campbell's action, at least as he reported it, was characteristically
disingenuous. 'To-day', he wrote on the evening of 26 May, when
the disembarkation was finished, 'I informed General Bertrand that,
in case either Napoleon himself or others might ascribe any under-
hand motive to my remaining here, I was ready to quit the island at
once, should such be his wish.' Bertrand's verbal reply, confirmed
in writing next day, was that Campbell's presence in Portoferraio
was 'indispensable', partly because so many British ships were visiting the island, partly because there was still a risk that Elba might
be attacked by Barbary corsairs and the Royal Navy was the only
means of providing the protection against such attack. For both

reasons, Bertrand concluded, 'I can but repeat to Colonel Campbell how much his person and his presence are agreeable to the Emperor Napoleon'.

The immediate advantages, it seemed to Napoleon, offset the unavoidable assumption that Campbell was either a spy, or would inevitably become one. For Campbell's continued presence, in the absence of any other diplomatic recognition, gave some standing to his government, implied a friendly attitude on the part of the British, and provided a link to the other Allies who had actually signed the treaty which had installed him on Elba.

By the late summer, indeed, the slow process of correspondence with London eventually brought Campbell formal authority to stay. When he received Bertrand's formal request to remain on Elba he put out in the brig *Swallow* in the hope of intercepting Pellew and Bentinck who were expected to pass close to the island on their way to Sicily, believing that these senior officers could endorse his actions, or give him new orders; but when he failed to make contact with them he asked Ussher to carry a letter to forward on to Lord Castlereagh, in which he copied Bertrand's note and again pointed out that he had received no instructions since he left Fontainebleau.

Two months later he received a reply, dated from the Foreign Office on 15 July, telling Campbell that his despatches had been placed before the Prince Regent, and hinting at the range of duties he was expected to perform without putting anything in precise or compromising terms. 'I am to desire that you will continue to consider yourself a British Resident in Elba', Castlereagh wrote, 'without assuming any further official character than that in which you are already received, and that you would pursue the same lines of conduct and communication with this Department which, I am happy to acquaint you, have already received His Highness's approbation.' By the time this letter reached him Campbell had already resigned himself to a long sojourn on the island, and to the fact that instructions so vague left him a worrying amount of discretion; only he, as the man on the spot, given responsibility but no authority, could decide how to implement Britain's ambiguous policy towards Napoleon. Each day, and with every fresh action, he would have to decide how to strike a balance between respect, supervision, and protection.

*

Although Napoleon was eager for the real though informal security offered by British ships in the harbour of Portoferraio, and by British control of the central Mediterranean, he knew that he must also make the most of his own resources. Over seven hundred men had arrived from Savona in the British transports. Most of them were thoroughly seasoned members of the Guard, picked men, for there had been intense competition for places; there were some fifty Polish lancers, a few mamelukes, some marines and gunners, a ten-man band and fourteen drummers. While this force might be sufficient for Cambronne to hold Portoferraio against a sudden attack there were no men to spare for duties elsewhere on the island. There was even a shortage of specialists. Some gunners from Dalesme's garrison were retained until more could be trained, and grenadiers from the Guard were put alongside miners from Rio to learn the essential skills of the sapper.

From the outset, therefore, Napoleon sought to expand his army by local recruitment. A handful of officers and men chose to leave Dalesme's regiments and enlist for service on Elba, and Napoleon made a special effort to recruit the Corsicans among them, for they were strong and brave, though ill-disciplined and prone to desertion; but even the promise of a special Corsican battalion, with bounties for recruitment, actually achieved very little. And while the recruiting officers whom Napoleon covertly despatched to Corsica and Italy attracted a number of volunteers, they were always running the risk of arrest by the French and Italian authorities who had no intention of letting Napoleon build up his power. The recruits they sent to Elba were never enough to keep the 'foreign' battalion at full strength and they were barely sufficient to offset losses from discharge and desertion.

The effort to make a show of defence around the island and to persuade Elbans to take a useful part in it was even less successful. Napoleon managed to maintain a small garrison at Longone, where the Spaniards had long ago built a strong citadel overlooking the harbour, and for some weeks he kept a group of soldiers busy with his scheme to fortify Pianosa. He patched up some of the watch-towers built over the centuries by Italian, Spanish, and French defenders, and by the Elbans themselves, but this was merely a temporary measure—the existing coastguards were so useless that Napoleon soon disbanded them, and the guns so old that before long he sold most of them off to Italy as scrap. The best he could quickly do for coastal defence was to site a few batteries in commanding positions and put some regular soldiers in charge of them.

He also set up a small and reasonably efficient corps of gendarmes which was responsible for maintaining domestic order and keeping a close watch on people travelling to and from the mainland, and about the island. All these activities, however, made only a marginal contribution to the security of Elba, and Napoleon was still dangerously short of men if a landing was made on the island. With the Guard and most of the Corsican battalion committed to the defence of the forts and harbour at Portoferraio, with Christian Lapi's poorly trained National Guard a doubtful asset for anything except parades and guard duties, he decided to replace it by a part-time Elban battalion, a sort of militia whose members could quickly be mobilized in an emergency. He told Drouot to form four companies, each with three officers, ten non-commissioned officers, eighty-seven men, and a drummer, and to arrange that each company should furnish a squad of twenty-five men to serve for one week every month—an ingenious means of keeping an active reserve while paying for only one quarter of the men at any one time. The Elbans were reluctant to become soldiers—Campbell noted that many had either sold or lost 'their arms, accoutrements, and clothing'—and cartoons of Napoleon playing at soldiers with straw men and wooden cannon were soon appearing in Paris and London. There was some point to them, for the plans for this 'Free Corps' produced such poor results that it was abandoned altogether by the end of the year.

Napoleon ran into the same kind of difficulty when he began to form his tiny navy. He needed a flotilla to run errands across to Genoa, Livorno, Civita Vecchia, and Naples, to keep an eye on trading vessels, and to provide some protection against pirates; but the Elbans preferred to sail about on their own affairs, or to go on fishing, and though Napoleon recruited sailors from Capraia and Genoa, and drafted in some marines from the Guard to stiffen discipline, his boats were always chronically undermanned.

The *Inconstant*, for instance, was a good 16-gun war brig of 300 tons, recently built at Livorno and armed with 18-pounder carronades, which needed about 100 men to sail and fight it efficiently; but Lieutenant Taillade, who commanded it with the help of a Corsican midshipman, could seldom count on two-thirds of the crew he would have needed if he had taken the *Inconstant* into action. There were similar shortages on the *Caroline*, a 26-ton Maltese esperonade, a half-decked and flat-bottomed craft which would easily be run up on a beach, and on the *Mouche* and the *Abeille*, two lateen-sailed feluccas which were taken over from the iron trade at Rio to patrol against smugglers and serve as general carriers around Elba. The *Etoile*, a fast and elegant three-masted xebec of 83 tons, carrying six 14-pounder guns, which Napoleon purchased at Livorno in August to assist the *Inconstant* in trade with the mainland, was also short of men. He was even hard pressed to find crews for the three small boats employed around Portoferraio harbour and in coastal waters—the *Ussher*, a ten-oared barge left by the departing captain, a gift and used by Napoleon himself, the six-oared *Hochard*, allocated to Bertrand and Drouot, and a smaller craft that was kept busy on general duties about the port. It was all a far cry from the great fleets at Toulon and Brest, and the line of battle which had once faced Nelson at Trafalgar.

★

Even such reduced military arrangements, however, were a significant burden for a man who was also in reduced financial circumstances, and when Napoleon made his first guess of what they would cost he told Drouot that he expected to spend up to a million francs on defence in the coming year. This was half the annual pension which had been assured to him by the new French government, and the scale on which he was establishing his royal household suggests that he had mentally committed much of the other half of that promised pension. It was a rash policy, for until the pension was paid Napoleon was entirely dependent on the reserve of nearly four million gold francs which Peyrusse had brought from France, on what he could collect from the iron mines, the tunny fishery, the salt flats, local taxes and customs duties, and on any other realizable assets he could claim as part of his royal domain; and the reserve was a dwindling source of funds which could not be replenished as it was used.

Napoleon's financial position was thus complicated and uncertain, reflecting his own ambiguous status on Elba as part pensioned emperor, part king, part landed proprietor, part commander-in-chief, and part civil governor. The matter was further confused by the way he charged expenses where it suited him, transferring money from one account to another, using vague headings for transactions he wished to conceal, and making no distinction between income and capital. It might be possible for the able Peyrusse to balance the books in the end, even if Louis XVIII were tardy or remiss in paying the pension, for experience had taught him how

Napoleon managed his money, but it would undoubtedly be months before Napoleon could really do more than speculate on his prospects.

At best, Campbell reckoned as he ran over Napoleon's likely receipts in a report to London, the King of Elba could expect to receive about 650,000 francs in a full year, and this estimate did not allow for the continuing effects of the blockade, a severe winter, and the dislocation caused by a change of regime. The iron mines accounted for most of this total, yielding a revenue of 500,000 francs, while the salt ponds (50,000), timber taxes (30,000), wine and oil duties (25,000) and the tunny fishing (24,000) were the only other significant items. Even if Napoleon made some tax changes, and trade improved, the island was too poor to carry a much heavier burden.

It was Napoleon's urgent need for a local income as much as his desire to impose an efficient government on the Elbans that made him move quickly to collect the taxes that had fallen into arrears in the last phase of the war. He justified the move as an advance on his French pension and at the same time sustained his pressure on Pons to hand over the outstanding money in the mining account and to concede that the future profits from the iron mines should be diverted from the Legion of Honour to the government of Elba. But this attempt to collect arrears of tax, Campbell noted, 'occasioned unusual outcry and supplications' which went on well into August; in Capoliveri, the hill-top community which had always enjoyed special privileges, resistance to the tax-collector became so fierce that Napoleon had to threaten to quarter a company of the Guard on the villagers, and Campbell noted a series of other incidents to show that there had been an abrupt change of mood. 'Napoleon appears to become more unpopular on the island every day', he remarked early in June, arguing that 'every act seems guided by avarice and a feeling of personal interest' and suggesting that 'the inhabitants perceive that none of his schemes tend to ameliorate their situation'. One of his gloomiest notes claimed that if Napoleon's oppression, injustice, and restlessness 'is not tempered by discretion, nothing but the military force of his Guards will prevent the inhabitants from rising against him'. Napoleon was clearly trying to do too much too quickly. The very small island, after all, had only very small resources, and its ruler would eventually have to come to terms with that fact.

8

THE FRUITS OF VICTORY

Out of sight, out of mind. In retrospect it seems astonishing that nobody in Paris gave much thought to Napoleon after his carriage had rolled away to the south, but an extraordinary war and an extraordinary victory were being followed by an extraordinary peace conference—a strange whirl of celebrations in which the conquerors and the conquered sat at the same dinner tables and shared the same gossip, and found it equally difficult to think seriously about the fallen emperor and the chaos left by the collapse of his empire. 'This capital is as bad as any other big city for business,' Prince von Wrede wrote candidly to the King of Bavaria on 6 May 1814. 'We eat, we drink, dance, see the sights and the women, but affairs do not move forward as one would desire.'

The Allied sovereigns, fortunately, were served by an oddly-assorted yet talented group of ministers who had learnt to understand each other when they worked together in the last stages of the war, and without their habit of compromise the coalition would have fallen apart before its members began to taste the fruits of victory. The Tsar's close advisers were the oddest of all, for none of them was a Russian. Count Nesselrode, a cautious man who was well aware of the difficulties created by Alexander's unstable moods and contradictory ambitions, was a professional diplomat of German descent who had been born in Lisbon; Carlo Pozzo di Borgo was a Corsican nationalist who harboured a long-standing animosity to Napoleon, and Giovanni Capo d'Istria was a Greek from Corfu who had entered the Russian service and become an expert on foreign affairs. The other ministers included Karl August Hardenberg, an energetic reformer who had done much to modernize the Prussian state and who bitterly resented Napoleon's treatment of his country; he had induced the vacillating Frederick William to join the final campaign against the Emperor. Clemens Metternich, a cultivated dilettant whose romantic conquests were said to include Napoleon's sister Caroline, was another expatriate professional, born in the Rhineland, who had become the calculating architect of Austrian policy. Lord Castlereagh was there because he was a parliamentary figure in his own right and Pitt's successor as the maker of great alliances against Napoleon. After taking much responsibility for the conduct of the war he was now hoping to strike a new balance of power and secure peace in Europe for a generation.

The ministers had little to do with the fate of Napoleon, for the Tsar had almost settled the terms of the Treaty of Fontainebleau before they all reached Paris. Their first real task was to draft the Treaty of Paris, which would settle the fate of France. 'The Allied sovereigns on the Continent stand in some fear of each other,' William Cobbett wrote in his *Political Register* on 11 June, noting their conflicts of interest. 'France does, and always will, hold the balance of Europe in her hands. Any one power joined with her must be more than a match for all the rest of Europe.' Napoleon himself had used that kind of leverage to break up the coalitions of his enemies, and the Allies were well aware that the game might be played all over again as soon as they began to deal with the fate of Poland, Saxony, and Italy. If, in the next few months, Talleyrand carried France towards Russia, he would thereby drive Prussia into alliance with Austria; if he courted Austria he would make Frederick William of Prussia even readier to collaborate with Tsar Alexander. The Allies were therefore eager to make a quick and easy peace while France was exhausted, with a million casualties, sacked cities, ravaged farms, and a mountain of debt to show the price of Napoleon's last campaign; and before the Treaty of Paris was signed they insisted on a secret clause which reserved to them the last word in the settlement of Europe. It was scarcely enforceable if they fell out with each other, but it offered them some assurance against the day when Talleyrand came back to the bargaining table on equal and potentially disruptive terms.

The most urgent and difficult task was to decide on the new boundaries of France, the issue on which Napoleon had fought to the end, trying to save the gains of the Revolution by standing on the 'natural' frontiers of the Rhine, the Alps, and the Pyrenees. Louis XVIII, who had been restored to the 'ancient' throne of the Bourbons, could more easily be persuaded to rule over the 'ancient' domains. He made an effort to claim more in Belgium, but this was the one point on which the British were adamant. As early as Christmas 1813 the Cabinet in London had decided on 'the absolute exclusion of France from any naval establishment on the Scheldt, and especially at Antwerp', as well as 'the restoration to Holland of her territory of 1792'; and Castlereagh was in no mood to argue with either the French or the Allies. 'After all we have done for the Continent in this great war', he told Lord Aberdeen at this time, 'they owe it to us and to themselves to extinguish this fruitful source of damage to us both.' When he came to Paris in the spring of 1814 he again insisted that Belgium was 'the source of so many wars' that neither the frontier fortresses nor the estuary of the Scheldt could be left in French hands, but he asked for nothing else. France could even have most of her captured colonies back in exchange for a French promise to abolish the slave trade within the next five years.

Castlereagh got what he wanted, and Louis was allowed some concessions in the Savoy Alps, but the Prussians were less successful with their demand for reparations in cash and kind. The other Allies thought them rapacious, arguing that they had profited enough from their systematic looting as they pushed into France and that such a burden could easily bring Louis down before he could establish himself in a country which scarcely remembered him, had no particular reason to like him, and was full of tensions bordering on civil war. The Allies were wise, as Castlereagh put it, to let France start again with 'clean scores', except for the claims of individuals whose property had been damaged or seized in the wars. They were equally sensible, in political matters, to keep their promise that the French should settle their own form of government, it being understood that Louis would make some gestures towards a constitutional system, and that he would find places for the soldiers and politicians who had made their careers under the Emperor. 'This is what I call a real peace,' Talleyrand wrote cheerfully to Metternich as the Austrian minister left for London. 'Don't tinker with it; don't alter it; if you do you will spoil it.'

That was easier said than done. Even this first round of negotiations had been protracted, and tiresome in detail, and it was merely a foretaste of the much more serious difficulties the Allies were to encounter as they tried to piece the remnants of Napoleon's empire into a new pattern. For the three states which dominated the centre of Europe were all driven by greed for new possessions and by a fierce rivalry which had scarcely been kept in bounds during the struggle against their common enemy. The Tsar wanted to expand his influence to the west, reviving Poland but making that much-divided country clearly subordinate to St Petersburg. The Austrians, whose rambling empire ran from southern Poland and Bohemia deep into the Balkans, wanted to keep Russia at a distance; they were equally eager to stop the Prussians annexing their neighbour Saxony and to prevent them advancing to the Rhine; and they were determined to make Italy little more than an Austrian protectorate by taking over the Tyrol, consolidating their hold on Lombardy and Venetia, and turning Napoleon's brother-in-law Joachim Murat off the throne of Naples. How could such jealous allies agree on a new map of Europe, or avoid falling into ruinous and warlike quarrels in the attempt to do so?

It was that question they were now to face in England.

★

On Monday 9 June, at 6.30 in the evening, the *Impregnable* brought the Tsar, the King of Prussia, and their suites into the harbour at Dover, where they stayed the night before going on to London. The whole cast of victors had decamped to Britain, accepting an invitation which dated back to the last days of the war when Castlereagh had too easily assumed that all would be settled by the summer and the Allied sovereigns would be free to join the Prince Regent in celebrating their common triumph.

There was one notable omission. Though Castlereagh had been keen, as he put it, 'to dilute the libation to Russia', so that the Tsar should not be made 'the sole feature for admiration', the Emperor Francis had excused himself and given Metternich full powers to represent him. He was no great loss. He was an unimaginative, intellectually limited man, torn between piety and a sensual nature which wore out his wives with annual pregnancies; he was dogmatic, fussy about trivialities, and little concerned with affairs of state, which he left to Metternich. Francis also knew that the British had no great liking for Napoleon's father-in-law or his army. 'Knowing as he does the feelings which prevail in England,' Lord Aberdeen wrote, 'I am not surprised that he should decline the risk of mortification.'

The Prince Regent was also risking humiliation when he drove through London, for his debts, his vanity, and his quarrel with his eccentric and estranged wife Princess Caroline had made him unpopular, and people hissed and hooted at him as he passed. He managed to drive a few miles down into Kent to welcome Alexander and Frederick William, but a ceremonial entry was out of the question and each man made his separate way into town, disappointing thousands who had lined the Dover Road to greet them. The Prince Regent even had to send an apology to Alexander, who was staying at the Pulteney Hotel in Piccadilly, explaining that he could not pay a courtesy visit to the hotel because 'he has been threatened with annoyance in the streets if he shows himself, and it is therefore impossible to come and see the Emperor'.

This was a poor start to a great festive occasion. After a cruelly cold winter, with the Thames frozen over, the coming of peace had coincided with a splendid spring and a warm summer which encouraged crowds to parade the decorated streets, cheering foreign dignitaries, especially the handsome young Tsar, who took every chance to contrast his popularity to the dismal reputation of his royal host. At the outset of the visit, indeed, both the Prince Regent and the Prime Minister, Lord Liverpool, had been well-disposed towards him, but his cool effrontery and his flirtations with the Whig opposition soon brought them round to Castlereagh's belief that Russia was a growing threat to British interests and that it would be unwise to let the Tsar dominate the coming settlement.

Castlereagh fretted all through the midsummer weeks, as the negotiations languished and the celebrations went on consuming time and money. Banquets, balls, illuminations, firework displays, and public events succeeded each other on a scale that seemed truly magnificent after twenty years of war, and was certainly in striking contrast to the gimcrack court which Napoleon was setting up in Portoferraio. One of the great reception rooms at Carlton House was itself big enough to contain the whole of the Villa Mulini; and a feast for seven hundred guests at the Guildhall, on 18 June, which was said to be 'as sumptuous as expense or skill could make it', cost £20,000, about a quarter of Napoleon's entire budget for a year on Elba. The climax to the jubilations was a grand naval review at Portsmouth on 23 and 24 June, when the foreign visitors, *The Times* wrote, were shown 'those tremendous armaments which . . . consecrated to Britain the domination of the seas'.

No one doubted the might of the Royal Navy, or the wealth of Britain—that combination of British powers had brought the Allies together to win the war; but it was not easy to see how they could be used to separate them in peace and to prevent their armies clashing as they argued over the spoils. Alexander's mischief-making, and the manner in which Metternich had adroitly turned it to Austria's advantage by flattering the Prince Regent and conciliating the Tory government, had shown Castlereagh how easily the balance between the former allies could change. All that had actually been achieved by 27 June, when the Tsar left for a visit to Russia before the peace congress resumed in Vienna, was a simple standstill agreement in which each of the Allies agreed to keep 75,000 men under arms until the map of Europe had been redrawn, and promised meanwhile that none of them would make any permanent changes in the parts of Europe they had temporarily occupied. The affairs of Poland, Saxony, and Italy would have to wait.

London was still in a festive mood as the sovereigns and their retinues slipped away, for on 1 August there was another great celebration to mark the jubilee of the Hanoverian succession. There were more fireworks, including a representation of the Battle of the Nile on the Serpentine, and all over the country there were beer-drinking contests and band concerts, cricket matches and dancing on the village greens. It was the greatest riot of pleasure the country had known since the Year of Victories in 1759. Yet all these feasts, as Castlereagh now realized, were haunted by the spectre of a new war as deadly and confusing as those which had been waged by the exiled King of Elba.

★

As soon as the celebrations were over Castlereagh left for Vienna. He knew that he would be away for several months, since the Congress was not due to begin until 1 October and he was aware that any gathering in Vienna was likely to last longer than expected; he would have preferred, he said, to meet in a city 'where business would be less liable to be interrupted by pleasure'. All the same, he hoped to be home by Christmas.

It was a vain hope. When the Allies decided to seek a comprehensive settlement in Vienna they said that they had a common desire 'for the good accord and understanding' necessary to establish a lasting peace, and Castlereagh was travelling out with full authority to seek what he described to Lord Liverpool as 'a just equilibrium in Europe' which would as far as possible be based upon the principle of legitimacy. But even before the Congress began it was clear that everyone had a different idea what that equilibrium should be, and that the pace of the Congress would be regulated less by the habit of past agreement than by the prospect of future tensions. 'While we were actually at war', Castlereagh wrote impatiently to Lord Liverpool in a letter complaining of the endless round of extravagant pleasures, 'events hurried even the most temporising to a decision; at present the irresolute and speculating have full scope to indulge their favourite game.'

How could there be a quick settlement when there were three huge armies sprawled across the Continent, and their masters were squabbling over huge tracts of territory? How could principle be balanced against expediency, though all the victors were strong for the restoration of legitimate governments, when Napoleon had so wrecked the royal houses of Europe? And how could the Congress proceed rationally when the three sovereigns who dominated it were despots, with all their whims, antagonisms, and ambitions compounding the problems of peacemaking? The Tsar, in particular, became so capricious that the gossips saw his erratic states of mind as the first signs of inherited insanity. On one day he would remind his allies that his Cossacks had ridden to Paris and could as easily ride to Vienna. On another he would show his spiteful distaste for the Bourbons, or the streak of quixotic honour which had led him to install Napoleon on Elba without thinking of the consequences, or the sexual jealousy that certainly spiced his relationship with Metternich. Uncertain whether he wanted people to fear or to love him, Alexander vacillated and everyone in Vienna waited uneasily. More than anyone else, as Castlereagh reminded him on 12 October, he could decide whether the Congress 'should prove a blessing to mankind or only exhibit a scene of discordant intrigue and a lawless scramble for power'.

The most difficult matters would have to be settled first, unfortunately, for the manoeuvrings about Poland and Saxony were to be a trial of strength and all the lesser issues would depend upon the outcome. Only then could the sovereigns and their ministers decide what to do about Norway, which the ambitious Bernadotte was seizing for Sweden with the connivance of the Tsar; about the status of the Pope and the Sultan of Turkey, about dynastic rights and successions in the smaller states, about diplomatic etiquette, the future of the slave trade, the control of such international rivers as the Rhine and the Danube, and the emancipation of the Jews. Every interested party in Europe flocked to Vienna—the Kings of Denmark, Bavaria, and Württemberg, Napoleon's stepson Eugene, and his sister Elisa, a scattering of princes and dukes, and even some representatives of German publishers concerned about the law of copyright. They all wanted to have their say.

They were given little chance to do so, for the great powers intended to make all the important decisions themselves and merely to promulgate them to the crowd of underlings who danced around them. The Congress was to hold only one formal session, in June 1815, when it was convened to approve the bargains which had been struck in a host of informal meetings during the previous seven months. The presence of so many claimants, however, gave Talleyrand his opportunity. By championing the rights of small states to a share in the proceedings, and by cleverly exploiting the differences among the former allies, he soon induced them to make France and Spain parties to the main discussions and eventually to admit him to their private discussions on more or less equal terms. His brilliant tactics thus changed the equation of power in Europe before the bargaining really began, and all through the autumn his influence steadily increased. By the end of the year he had convinced Metternich that a new alliance with France was the only means of resisting Russian pressure, weaned Castlereagh from his original support for the Prussians, and persuaded him that a combination with France and Austria offered the only hope of preserving peace.

For it was a general and enduring peace, rather than any particular advantage, that Castlereagh most wanted at Vienna. His country was exhausted by the struggle against Napoleon, and by the still unfinished war against the United States. The Whig opposition in the House of Commons was strenuously opposed to any more fighting and more high taxes, critical of the Bourbons, hostile to the Austrians, and inclined to support both the Prussians and the Tsar, and his government was so shaky that, in the search for trustworthy friends, it had turned away from its old allies and towards France. 'The more I hear of the different courts of Europe, the more I am convinced that the King of France ... is the only sovereign in whom we can have any real confidence,' Lord Liverpool told the Duke of Wellington shortly before Christmas 1814. 'The Emperor of Russia is profligate from vanity and self-sufficiency, if not from principle. The King of Prussia may be a well-meaning man, but he is the dupe of the Emperor of Russia. The Emperor of Austria I believe to be an honest man, but he has a minister [Metternich] in whom no one can trust, who considers all policy as consisting in *finesse* and trick, and who has got his government and himself into more difficulties by his devices than could have occurred from a plain course of cheating.' Castlereagh was equally sceptical about his colleagues. 'I beg you will not give any money at present to any of the Continental Powers,' he wrote to

Lord Bathurst on 30 January. 'The poorer they are kept, the better. It prevents them from quarrelling. Time enough to settle
accounts when we know who deserves it.'

The setting for these complex and protracted negotiations, in
which suprisingly little thought was given to Napoleon's continuing
capacity to influence events, was the most brilliant social event in
the history of Europe. Francis himself had to accommodate another
emperor, four kings, one queen, two hereditary princes, three
princes of the blood, and three grand duchesses; and outside the
Hofburg palace more than two hundred heads of princely and ducal
houses had to be found apartments, fitted into the most complicated pattern of protocol ever devised, and sustained by crowds of
courtiers and flunkeys. The Festivals Committee, set up to organize
the round of gaiety, had done its work with a will and the visiting
notabilities were regaled to the point of distraction by balls, banquets, tournaments, theatricals, balloon ascents, autumn picnics
and winter sledging parties in the Auergarten, and a performance of
Fidelio conducted by Beethoven himself. There was a pleasantry
current in Vienna which said that the Tsar loved for everyone, the
King of Prussia thought for everyone, that the King of Denmark
talked, the King of Bavaria drank, the King of Württemberg ate,
and the Emperor of Austria paid for everyone. It was said that
Francis lavished over thirty million florins on entertainments at a
time when he could not pay his army and even Castlereagh, who
was a man of modest tastes, spent £15,000 in four months and got
through a private stock of over ten thousand bottles of wine.

It was the last great party of the crowned heads of Europe, reas-
serting themselves against the populist tide which had flowed across
their countries in the wake of the French armies, and each glitter-
ing occasion was alive with intrigue. 'All the courts and missions
are deeply engaged in spying on each other', Prince Starhemberg
said, and the busiest spy of all was Baron Hager, the head of the
Austrian secret police, who gave the Emperor Francis a daily report
on the tittle-tattle of the town, the private letters opened in the post
office, the contents of waste-paper baskets, and the official despatches to and from the foreign delegations which had been
opened by his agents. In this climate every secret grew into a
rumour within the hour, and had flowered into an accepted fact
within a day or two; and though much of the gossip was trivial or
misleading, the talebearers generally caught the changes of mood
and the drift of policy.

'Napoleon in the Isle of Elba has in this case only to be patient,' General Sir Robert Wilson wrote in his diary soon after the Allies had sent the Emperor into exile: 'his enemies will be his best champions.' Before the year was out his prediction would come true, and at Vienna they would be stumbling towards a new war which none of them wanted, which none of them could afford, for which none of them was prepared, and from which only Napoleon could benefit.

9

FAMILY AFFAIRS

As May passed into the blazing heat of June, and a fatiguing siroc-
co blew out of Africa, Colonel Campbell felt the need for 'release
from the sultry confinement of Elba', and he began to plan a long
furlough in Italy. He was bored, for the novelty of his situation was
wearing off and Portoferraio had few diversions beyond the little
court in the Villa Mulini, and he complained of the 'wearisome
feeling' brought on by the effort to write up his journal, to draft and
copy his despatches, and to deal with 'an immensity of correspondence in a public way'. He was still suffering from his wounds,
which had left him rather deaf, and weak in the back, and he hoped
to benefit from the livelier company and the treatment he would
find at the thermal baths at Lucca. And he felt he could combine
pleasure with duty by going to Florence and Rome. Napoleon's
'restlessness', he explained, 'made me anxious to compare my
suspicions with what information I could obtain on the Continent'.

There was no overt reason for Campbell's suspicions. On the
contrary, Napoleon continued to behave like a man resigned to his
fate and more troubled by family affairs than by the fate of nations.

On 27 May, for instance, Campbell wondered whether he had
privately heard that his wife and son had either abandoned him, or
were being held in Vienna. 'He has not made any such arrangements as evinced any expectation of his being joined by Marie-
Louise', Campbell wrote, 'nor has he mentioned her in any way.' It
was about this time, moreover, that the Emperor received his last
letter from the former Empress Josephine, writing from Malmaison
shortly before she died of diphtheria on 29 May. She was still much
attached to him, and she even offered to join him in Elba once the
fall of his dynasty removed the reason for his divorce on the
grounds of her childlessness, and she realized that Marie-Louise
had in fact deserted him.

'It is only today that I can calculate the full extent of the misfortune of seeing my union with you severed by law,' she wrote in her
undated letter. 'It is not for the loss of a throne that I mourn for
you . . . for that there is consolation, but I am desolated at the grief
you must have felt in separating yourself from your ancient

companions in glory . . . . You must also have wept at the ingratitude and desertion of friends on whom you thought you could depend. Ah! Sire, that I cannot fly to you to give you the assurance that exile can only frighten vulgar minds, and that, far from diminishing a sincere attachment, Misfortune lends to it a renewed force.'

On the day that Josephine died in Paris the new King of Elba was celebrating the Feast of San Cristino, the patron saint of Portoferraio, which served as a suitable festival to mark his assumption of the throne. There was an official Mass in the morning, to which he rode in his state carriage down a route lined by men of the Guard, and in the evening there was a reception organized by the town council. Two days later the Neapolitan frigate *Letizia*, named after Napoleon's mother, brought his sister Pauline into Portoferraio. She had been expected for the festivities but she had kept the vessel lingering off Villefranche for eleven days while she waited for the weather to give her an agreeable crossing. She did not like the Vantini house, which was being refurbished for her, and stayed at the Villa Mulini; while she was there she gave Napoleon a diamond necklace to buy a small estate three miles inland at San Martino. This was to be her property, though he was to have the use of it as a convenient retreat from the disagreeable heat and odours of Portoferraio, and she may even have expected to use the house as her own residence when she returned from an extended visit to her sister Caroline in Naples.

On this first visit, however, Pauline was clearly in a hurry, for within two days she had sailed again for Naples; despite her love of parties, she did not even wait for 'an impromptu fête' given by the officers of the frigate *Curaçoa* and the brig *Swallow* to mark the birthday of George III on 4 June. It was certainly not a nice sense of protocol that made her leave, for the British officers were at pains to avoid giving offence to anyone. They flew the Royal Standard at the mainmast, the white flag of the Bourbons at the foremast, and Napoleon's new Elban ensign at the mizzen, while the *Dryade* and the *Inconstant* flew the British colours. All the presentable inhabitants of Portoferraio were invited; General Dalesme and the officers of the retiring French garrison attended in company with Drouot, Bertrand, Cambronne, and officers of the Guard; and about nine in the evening, rather to everyone's surprise, Napoleon himself came off in his barge and made himself agreeable to the notabilities on the quarterdeck.

★

Superficially, all seemed to be going well. Yet Campbell was in such a suspicious state of mind that he saw the possibility of danger in everything—in the departure of the French and English frigates, in the replacement of the French garrison by Napoleon's own loyal Guard, and in Pauline's fleeting visit, for he was convinced that she had brought secret reports from Bonapartist agents in France and Italy and that she had hastened on to Naples with conspiratorial messages from Napoleon to his brother-in-law Murat. He even suspected that the peculiar episode of the lazaretto, which was probably caused by Napoleon blundering into a squabble with the sanitary authorities in Livorno, might actually have been staged to give Napoleon an excuse to stop movement in and out of Elba while he considered a sudden descent on Italy.

This peculiar dispute began when Napoleon suddenly decided to assert his sovereign rights by setting up his own quarantine station in the harbour at Portoferraio, and then insisting that all vessels coming to Elba should use it instead of getting the customary clearance at the lazaretto in Livorno. The authorities in the Tuscan port retaliated by putting a twenty-five day ban on all ships which touched at Portoferraio, and other ports in the Mediterranean soon followed suit. This damaging decision greatly annoyed Napoleon, who was convinced that it arose 'from jealousy of his projects, and from commercial intrigues' on the part of the merchants in Livorno; 'he persists in refusing compliance to long-standing practice', Campbell noted, 'though it cuts off his communication with every other part of the world, except by clandestine means, to his own loss and inconvenience'.

These difficulties delayed but did not prevent Campbell's departure—he had, he said cryptically, 'obtained an opportunity of evading the quarantine'; and his suspicions that they might be connected with some revolutionary intrigue in Italy made him all the more anxious to go to Livorno (as he delicately put it) 'for the security at once of my correspondence and for information', and then to go on to Lucca and Florence. He felt out of touch, for there were no accredited British diplomats in Italy at this time and he had to turn to men who worked in the shadow zone where it was hard to distinguish between trade, intelligence, and diplomacy. In Livorno he saw Grant, who had once worked with a network of agents based at Genoa, and had more recently been vice-consul in the Tuscan port; he also met Felton, who had succeeded Grant; and in Rome he talked to Robert Fagan, who had been engaged on a number of mysterious missions for Castlereagh. The Austrians were helpful too. When Campbell reached Florence he was received by Count von Starhemberg, the Austrian commander in Tuscany, who was 'extremely frank' and agreed to pass on any reports he received from his spies on Elba.

Campbell was glad of any help of this kind. Apart from employing a man named Ricci, who was to act as temporary vice-consul in Portoferraio and to pass on the gossip of the town, he seems to have made no arrangements for watching Napoleon while he was away from the island. But his journal and his despatches to London show that he did not learn much from Ricci or Starhemberg's men. Napoleon 'continues the same sort of life as before', he concluded from their scrappy evidence, 'engaged in perpetual exercise, and busy with projects of building, which, however, are not put into execution'. More seriously, according to Starhemberg, Bonaparte was carrying on a constant exchange of letters with his relatives. Even this was not such exciting news as it seemed. When the Austrian authorities opened the letters they disappointingly proved 'quite innocent', and Campbell had nothing more to report than the Count's conviction that 'they were merely sent to blind their other correspondence, carried on through more direct and clandestine channels'.

Campbell was clearly enjoying the change of scene, and the smart company he found in Florence. Among his new acquaintances, moreover, was a lady whose notable charms made the city seem much more attractive than Portoferraio, and he was so keen to stay there that, early in July, he wrote to Castlereagh asking for a diplomatic appointment in Tuscany and permission to restrict his duties in Elba to an occasional brief visit. Yet the pleasures of life in Italy did not wholly dull Campbell's sense of unease, though he was glad of an excuse to continue his surveillance of Napoleon from the more congenial side of the Piombino Channel. 'As his schemes begin to connect themselves so openly with the neighbouring Continent, and my information from Elba is so very detailed and correct', he wrote, as if to persuade himself, 'I think the spirit of my duties will be better fulfilled by not shutting myself up in quarantine.' He therefore went to Rome for a fortnight and had two audiences with the Pope, who expressed 'great uneasiness at Bonaparte being so near to Italy' and made him feel even more doubtful about Murat's intentions.

*

Italy was a name, and Italian a language, but the people who lived in the peninsula had been divided since the fall of the Roman Empire. The Italians, said Lord William Bentinck, had long been 'despicable slaves of a set of miserable petty princes', and all through the latter part of the war he had schemed to create a united Italy under British patronage and to make it 'a powerful barrier against Austria and France'. He had even taken troops under his command to occupy Livorno and to revive the Genoan republic. But his own government had repudiated him. 'It is not an insurrection we now want in Italy,' Castlereagh wrote to him on 3 April 1814. 'We want disciplined force under sovereigns we can trust.'

Sicily was ruled by Ferdinand, the Spanish Bourbon who was bent on recovering the other half of his kingdom, still occupied by Murat and Napoleon's sister Caroline, and everyone knew that Murat's position was precarious. Britain had never recognized him; the Austrians distrusted him, though they had made a still unratified treaty with him when he deserted Napoleon towards the end of the war; and the Bourbons in Paris were desperately eager to see the back of him. The Papal States in the centre of the country had been disorganized by a succession of occupations, and the Pope himself—who had been Napoleon's prisoner at Fontainebleau for the last five years—had returned in penury. When Bentinck met him on the way back to Rome he remarked that the poor man's wardrobe was down to four shirts, and he lent him 4,000 crowns of British secret service money to help set him up again in the Holy City. In the north, moreover, where national sentiment was strongest, the Austrians had occupied Venetia, Lombardy, Piedmont, and Tuscany, and secret clauses in the Treaty of Paris had permitted them to annex Venice and Milan.

The peculiar relationship between Napoleon and Joachim Murat was a further complication; whether they were on good terms or bad, whether they were in secret agreement or political rivals, they were brothers-in-law, they were near neighbours, their kingdoms were the last relics of the Bonaparte dynasty, and in the end they would stand or fall together.

Each of them realized this fact, and neither of them liked it. Murat had long resented the way in which Napoleon had treated him as a mere vassal, running Naples largely for the convenience of France, and when it came to the point he was prepared to abandon the Emperor to save his own throne. So was his wife, who had no intention of being ruined because her brother insisted on fighting to the bitter end. 'What's our aim?' Caroline had asked Murat as early as 1811. 'It's to maintain ourselves where we are and to keep the kingdom.' She was a strong-willed and intelligent woman, who had since done her best to stay on good terms with the Austrians—and especially with Metternich, who had once been her lover—but it had not been easy to manage her husband. As Napoleon knew very well he was a brilliant cavalry commander, much like Ney, and like Ney he was susceptible to flattery, politically incapable, and driven by an unstable ambition. 'He is more feeble than a woman or a monk when he doesn't see the enemy,' Napoleon once said. 'He lacks moral courage.'

Such temperamental differences might be overcome. Napoleon might even find it convenient to forgive Murat's defection. But in the autumn of 1814 each man had good reasons to play his own game, and to keep a formal distance. Even family correspondence was reduced to a minimum, and Murat was so keen to show that he had cut all links with Napoleon that he refused to ship wheat and cattle to Elba. For he knew that his survival depended on the general goodwill of the Allies and the particular protection of the Austrians. The Emperor Francis was known to dislike him. 'I hope the King of Naples will be the architect of his own ruin', he said to a French diplomat in July 1814. He also suspected that if Austria were ever in difficulties Murat would seize his chance to drive them out of Italy and set himself up as king of the whole country. But he felt obliged to stand by the 1814 treaty which recognized the turn-coat as King of Naples, at least until he saw which way the wind was blowing at the Congress in Vienna, or Murat himself did something foolish; and in the meantime Murat kept the Bourbons out of the Italian peninsula and provided a useful counter for Metternich to use in bargaining with the French.

Napoleon was equally anxious to avoid anything that looked like collusion between Elba and Naples. He not only feared that the Allies would seize any excuse to transport him to a more distant and safer place of exile, he was also afraid that Murat might do something impetuous that would compromise them both, and he was reluctant to let his unstable brother-in-law put himself at the head of the movement for Italian unity.

This was the most pressing question of all, given the divided and profoundly unsettled state of Italy. The country seemed to be on the verge of insurrection, but the nationalist movement needed a popular and militarily experienced leader if it was to have the slightest chance of success. Murat was one candidate. Eugene was another, even though he had given up his command and left Italy for Bavaria in late April. Napoleon was the most attractive of all. For his family originally came from Italy. He spoke Italian. He had twice driven the Austrians out of Italy in a spectacular campaign, and entered Milan as a liberator; and even though he had later disappointed the Italian liberals he had at least made his son King of Rome and his stepson Eugene the Viceroy of Italy.

Among many clandestine messages that passed between Italy and Elba that summer was a stirring appeal from a group of revolutionaries, said to include men of standing in Corsica, Genoa, Piedmont, Milan, the Papal States, and the Two Sicilies, and to have the backing of wealthy bankers in Genoa. 'You have shown an exhausted world what you can do with your sword,' they wrote in a letter supposedly carried over by Cambronne from Savona. 'Finish by showing what your genius can do as a lawgiver and citizen king ... put yourself at the head of European civilization.' This proposal was bound to tempt Napoleon, for their plan for driving the Austrians out of Italy was very like one of the schemes he had considered in the last desperate days of Fontainebleau, when he had proposed to march the remnants of his army south, to strike over the Alps to meet Eugene, and then to offer Austria the choice of a new war or a settlement that would make him master of Italy and the ally of his father-in-law for the third time. Yet he was well aware that it is easier to hatch conspiracies than to succeed in them, and the complications and risks of a sudden descent on Italy were so great that—despite the fears that Campbell kept reporting to London—he clearly preferred to play a waiting game. All the same, the conspirators had made one telling point. 'As soon as the moment comes for the Emperor Napoleon to leave his island', they wrote with remarkable prevision, 'he should make the crossing by the shortest possible route, to prevent the news of his departure preceding his arrival. It is essential that his return should produce on his enemies the effect of the head of Medusa, and that he should appear to the soldiers like the god of victory fallen among them from the sky. The greater the surprise the greater the magic of the enthusiasm.'

Campbell did not really know what his government expected him to do in this confused situation, and he was understandably anxious to keep all British dealings with Napoleon in his own hands. But lack of clear authority and poor communications combined to frustrate him. He had no control over the British warships still operating along the Italian coast, and he was disturbed to find frigate captains such as Ussher and Tower taking it upon themselves to fetch and carry for Napoleon without specific permission or any reference to him. While he was in Florence, for instance, he was surprised to see two officers from the *Undaunted*, who had been sent to ask when Marie-Louise was expected, and to discover that Ussher had put into Livorno in the hope of meeting her and taking her over to Portoferraio. Soon afterwards Tower suddenly arrived at Civita Vecchia in the *Curaçoa*, bringing an invitation from Napoleon to his mother in Rome and offering, after he had paid a brief social call on Murat and Caroline in Naples, to take Madame Mère and her household to join her son.

All this hobnobbing with the Bonapartes upset Campbell, although it was due to muddle rather than mischief. Both Ussher and Tower had received a delayed and vague message from General Köller, asking for the use of a British ship if Marie-Louise turned up in Livorno on her way to Elba. Since this request had been transmitted by Admiral Pellew both captains clearly assumed that no one would object if they made themselves useful as ferrymen to Napoleon's relatives. It was a common courtesy for gentlefolk, let alone royalty.

But Campbell did object. He sent Towers a tart rebuke and wrote formally to Castlereagh to insist that all approaches to Napoleon and members of his family must be made through him. He also saw to the removal of Napoleon's mother himself. The *Curaçoa* had gone off on other duties, her captain's letters suggesting that he was chastened but unrepentant, and as there were reports of corsairs off Civita Vecchia it seemed safer for Madame Mère to go overland to Livorno and cross from there. Campbell himself left from Civita Vecchia in the *Swallow* on 23 July, called at Elba to tell Bertrand that Madame Mère was coming — he was unable to land because the quarantine edict remained in force until the end of the month — and went on to Livorno to wait for her arrival on 29 July.

Madame Mère, Campbell noted, was 'very pleasant and unaffected' and they had an agreeable crossing on the *Grasshopper* brig, which was replacing the *Swallow*, Madame talking freely to Captain Battersby about her famous son. When the brig reached Elba, however, they were met only by a valet, who told them that Napoleon had expected his mother on the previous day and that he was now out in the country. At this news, Campbell remarked, 'she

seemed greatly agitated and mortified', and she refused to land until Drouot and Bertrand had been hastily summoned to turn out a
guard of honour, line the streets with cheering citizens, and generally appease her chagrin. That night she was installed in the Vantini house, originally intended for Pauline and now turned into something like a dower house close to the Villa Mulini.

She was sixty-four. Though she was small, she had an imperious
manner. 'I have seen eminent people more intimidated in front of
her than in front of the Emperor', Pons de l'Hérault remarked, and
Napoleon himself usually treated her with respectful deference. 'My
mother could govern a kingdom,' he once said; 'she has excellent
judgement, and never makes a mistake'. She was the only person,
one of their Corsican acquaintances recalled, who could make him
listen, or 'force him to account for his occasionally bizarre and ex-
travagantly original behaviour'. They had their quarrels, usually
about money or the status accorded her as the Emperor's mother, but
such disagreements were a commonplace in the Bonaparte family,
who were all greedy for riches and place; and at least Madame
Mère was less extravagant than her children. By the time of the collapse in 1814 she had saved about three million francs from her
civil list income, and her jewels and other property may have been
worth half as much again. She had come a long way since she had
fled from Corsica as a penniless widow with eight children to take
refuge in revolutionary France, and yet she had changed very little.
'Madame was a Corsican in the full sense of the word,' Pons said.
'Her speech, her manners, her memories—everything evoked her
early years, and one sometimes asked oneself if she had ever left
Corsica.' From the moment she arrived on Elba, indeed, she devoted herself to the Corsicans on the island, soliciting posts and
other favours for them so brazenly that Napoleon eventually declared that he could not afford his mother's generosity to her compatriots, and told Bertrand to put a stop to it. He was always careful to conduct such business through intermediaries, for he was
curiously irresolute when he had to reprove any member of his
family face to face, and he was particularly susceptible to the combination of harsh pride and possessiveness with which his mother
had treated him since childhood.

All the same, he seemed genuinely glad of her company, and her
presence certainly helped to create the desired impression of settled
domesticity. Her house was tiny by comparison with her Paris residence, which was big enough to be converted into the Ministry of

War when she sold it. Her household was small—her chamberlain, Colonna d'Istria, a retired French general, a pair of ladies-in-waiting, and a reader; and she lived very plainly, seemingly content to revert to something like the style she had known when she was a young woman of good family in Ajaccio. Yet Napoleon was more dutifully attentive than he had been in Paris. He went to see her every day, invited her to spend her evenings at the Villa Mulini listening to music or playing *écarté* and *reversi*, and expected the guests at his Sunday morning *levée* to pay court to her before the whole party went to Mass in the town church. Even if the Bonapartes had come down in the world the mother of them all was to be given her due.

★

Campbell had so far enjoyed easy access to the Emperor and had been his companion on many rides about the island, but as soon as he returned to Elba at the beginning of August he felt that there had been a significant change in Napoleon's attitude towards him. On the evening of 3 August, he noted irritably in his journal, he and Captain Battersby were summoned to the Villa Mulini, halted by the guard, and sent to wait in an ante-room while Napoleon finished a card game with his mother and Bertrand. After they had cooled their heels long enough to arouse Campbell's prickly temper they started to leave, and at this sign of resentment they were immediately shown into the salon, where the Emperor 'came up tripping and smiling' and as full of questions as ever. Had Campbell seen the Pope? What was happening in Sicily? Was it true that the Prince Regent's daughter had run off in a hackney-coach to escape from her father? When the queries flagged Campbell reported that he now had the Prince Regent's permission to remain on Elba and he noted that Napoleon, who had previously seemed eager for him to stay on Elba, 'only nodded and said "Ah! Ah!"' Campbell clearly felt rebuffed, and in this petulant mood, giving no other reason than his desire to discuss the matter with Admiral Hallowell who had taken over command in the Mediterranean, he decided to go back to Livorno when the *Grasshopper* sailed next day.

When Admiral Hallowell arrived in Livorno on 20 August Campbell was still harping on his grievance about Captain Tower and fussing about other ships which had put into Elba to gratify their captain's desire to see the Emperor. He would have been wiser to let such casual arrivals reinforce his own haphazard surveillance of

Napoleon, for both his journal and his despatches at this time contained nothing better than snippets of gossip, but he pressed his point of protocol. Since Hallowell had no instructions from the Admiralty to maintain a close naval watch on Elba he conceded it, criticizing 'several instances of voluntary court and unnecessary visits paid by naval officers at Portoferraio' and promising to continue Pellew's policy of 'leaving a man-of-war upon the station in case of any extraordinary circumstances'. The warship thus placed at Campbell's disposal had no orders to patrol the Piombino Channel, or to search for signs of clandestine activity, but she none the less seemed like a guard-vessel and her presence could only enhance the misleading impression that the British navy was keeping the King of Elba safe on his island.

This ambiguity was eventually to prove as unfortunate as the uncertainty about Campbell's personal status at the Emperor's court, and it also stemmed from Castlereagh's refusal to take formal responsibility for Napoleon while allowing British officers to become practically involved in his affairs. Castlereagh, indeed, was even trying to save money on petty details. In a letter which reached Livorno on 29 August he asked for duplicates of Campbell's reports to be sent to Vienna, where he would be attending the peace congress, and reminded Campbell that he must only go to the expense of couriers on occasions which 'require immediate despatch'. Campbell might fidget about Napoleon's intrigues, but since he spent most of August and September in Italy he obviously did not feel any urgent need to observe them at first hand or to trouble Castlereagh with special messengers. The first eight days of September were covered by a single laconic entry in his journal. 'Florence and Baths of Lucca.'

*

All through the summer Napoleon kept himself in the public eye and busy with proposed improvements to his island estate. His birthday, for instance, was celebrated as a public holiday, and the festivities not only ran the municipality of Portoferraio into debt but put an additional strain on the citizens who formed part of the little court circle. 'There were not six families in Portoferraio who could afford the exceptionally splendid dresses their wives required for the occasion', Pons recalled, and he gave a jubilant account of the dancing in a wooden ballroom erected in the Piazza d'Armi for the use of the officers and gentry, of the first race-meeting on Elba, with horses ferried over from Italy, of royal salutes from the forts, a *levée* for Napoleon and another for his mother, an official Mass, fireworks and illuminations. Though Pons was still conducting his protracted dispute with the Emperor about the accumulated funds from the iron mines he made a great effort for the occasion, and to his delight he won the prize for the finest display in Portoferraio. Each of his eight windows contained a letter, with a star above it, spelling the name Napoleon, and there was a twenty-foot Anchor of Hope in the centre, the whole design being set out with red, white, and blue lamps. Campbell later complained that his allegory was alarmingly explicit, but Pons insisted that it was nothing but an innocent compliment. 'I knew nothing of the projects of the Emperor,' he declared; 'in fact, I do not believe that the Emperor had any projects at that time.'

It certainly seemed from appearances that Napoleon still had nothing currently in mind except the modernization of Elba. He copied French rules and regulations, down to such details as the length of the hunting season. He told Balbiani to enforce the collection of taxes and generally to raise the standard of administration on the island, and almost every day he issued some new order to Bertrand or Drouot: to provide fresh water supplies and better sanitation, to stop animals wandering untended in the lanes of Portoferraio, to harry traders who left putrid fish and vegetables in the gutters, to limit prostitution, to ensure that the soldiers had a diet which would ward off scurvy, to establish an efficient hospital, to upgrade the rough tracks between the villages, and to encourage the people to plant trees instead of cutting them down. They were all seemingly sensible orders, yet even Bertrand came to the conclusion that the flurry of activity in the summer of 1814 was really a means of releasing the energy which had once driven the machinery of government all over Europe. 'His Majesty', Bertrand later recalled, 'dictated letters about fowls, ducks, meat and all eatables as if he was dealing at Paris with matters of the greatest importance.'

There were, of course, other reasons for starting and abandoning projects. Some were tactical. In the early days on Elba it was in Napoleon's interest to suggest that he was turning the island into a model and peaceful principality; and in the autumn, when he wanted to pillory Louis for failing to pay his pension, he began to plead poverty as an excuse for retrenchment. And some were temperamental. Naturally restless, and a prey to whims, Napoleon easily became bored with a scheme when the novelty wore off, or it proved too difficult or costly.

The Pianosa project was a case in point. Soon after Napoleon visited the island in May he decided to make it into a Utopian colony for deserving Elbans, and he began with characteristic enthusiasm. He intended to settle each family on its own plot of land, with two cows and ten sheep, and mulberry trees to mark the boundaries between the plots. He was going to plant Black Forest acorns for oaks, pines on the poorer land, fruit trees, vines, grow much of the wheat that Elba had currently to import from Italy, and start a stud for breeding fine horses. He even appointed the parish priest of Campo to the cure of the as yet unsettled souls, and started negotiations with a man in Genoa who was to manage the whole scheme under contract.

While these ambitious plans were being drafted Napoleon sent out Major Gottman from Longone with a small party which was to put the place in a state of defence and build a barrack. It was a disagreeable duty, for the men had to live in a cave; they were to begin, Napoleon told them with his customary passion for detail, 'by making a fire to cleanse it of insects'. They were forbidden to hunt or to cut firewood; they were put on naval rations of salted meat, biscuit, rice and wine; and they were worked very hard, for Napoleon was eager to get on with his model village. On 16 June he was already asking for plans to show where the church, the store, and the commandant's house would be sited.

But nothing had gone right on the island, and this house had become a particular source of trouble. Gottman was a boorish and self-important man who had insisted that his own house should be built before the barrack, and this had led to such a fierce quarrel with his lieutenant that eventually Napoleon himself had to go over to dismiss Gottman and to prevent the two men fighting a duel. By then his enthusiasm had wilted right away. He had not only run into difficulties with the contractor in Genoa, but there were rumours of plots to abduct him during a visit to the island, and after three visits he stopped going there. The whole ambitious scheme was now abandoned; the island was used simply as a grazing ground for the horses of the Polish lancers, and as a kind of penal settlement for military defaulters.

When he stayed at Longone in September, moreover, Napoleon started another set of projects that came to little or nothing. He began to convert part of the old Spanish fortress into a royal residence though he soon changed his mind and decided instead to enlarge the house which Pons occupied as director of the mines at Rio; he devised a new harbour to protect the ships loading with ore; and he

then had an idea to avoid the expensive shipment of ore by building iron furnaces at Rio, although there was no coal on the island,
no wood for charcoal, and no one with the necessary skills. Within
weeks these projects too were forgotten.

But the construction of the house at San Martino was the most
striking example of sudden interest followed by indifference. When
Pauline bought the property it was only a simple farmhourse set in
a few acres of wooded hillside with a splendid view to Portoferraio
and the sea, and Napoleon immediately told the sculptor Paolo Bar-
gigli, originally brought from Carrara to set up a school of art, to
turn architect and make the building into a rural copy of the Villa
Mulini, with plain pinkish walls, green shutters, and a red-tiled
roof. Although Napoleon added an upper storey to take advantage
of the slope, leaving the lower floor for a bathroom and accommodation for the staff, his own quarters were even more cramped
than they were in Portoferraio. His bedroom with a fine view down
the valley, was fourteen feet by eleven, and its walls were painted
to give the illusion of a tent. The study, at the other end of the
suite, was the same size and between these two rooms, on the seaward side of the house, was a salon, twenty-eight feet long and sixteen deep, with two doves painted on the ceiling—supposedly to
represent the Emperor and Empress. Behind the bedroom was a
room for a valet and an ante-chamber; behind the study were two
other small rooms, which could be used by Bertrand and Drouot.
At the rear, and corresponding to the large upstairs salon in the
Villa Mulini, was the Egyptian room, almost twenty-eight feet
square, paved in marble with a sunken basin for a fountain fed by
the spring which rose above the house. The walls of this extraordinary chamber were decorated with paintings of camels, palms,
minarets, and Mameluke cavalry, with a freize of hieroglyphs and a
large zodiac on the ceiling.

Napoleon had no sooner approved the design for this country retreat than he sent twenty masons and some of his Guard to work on
it, and every day he went out to watch and encourage them as they
altered the house, built a new drive, saw to the crops on the additional farmland he had bought, and generally put the place in
habitable order. But the work lagged, and though he had pressed
Bertrand to have the bedroom, salon, and study finished by the end
of July he had to sleep in his tent when he spent a few days there
at the beginning of August. Then he seemed to lose interest,
though he let the workmen finish the house; and during the re-

mainder of his stay on Elba he seldom went there except to take advantage of the carriage road for his daily drive. Nor did Pauline choose to occupy it when she returned to Elba. The house simply stood empty, a relic of an abandoned plan. On one of the columns in the Egyptian room someone—it was said to be the Emperor himself—had scribbled the words *Ubicumque felix Napoleo*: Napoleon can be happy anywhere. But not, it seemed, in San Martino.

10

TWO KINDS OF LOVE

A few days before the end of July Napoleon suddenly announced
that he wanted to set up a summer camp in the hills. He gave good
excuses for the move—the hot weather, the noise and inconvenience of the building work at the Mulini, and the growing number
of tourists in Portoferraio, saying tetchily that 'they come to look at
me as if I were a wild beast.' But there was an undisclosed reason
as well. He was expecting a visitor sometime in August, and he
needed a secluded place for the rendezvous.

He soon found what he wanted at the west end of the island,
where there was a small mountain chapel devoted to the Madonna
del Monte, and a disused hermitage with four cells. The site had
many attractions. It was cool, shaded by chestnut trees, scented
with the rosemary and thyme which reminded Napoleon of his Corsican childhood, and blessed with a splendid view across to Corsica
itself. On a very clear day, with a glass, a man could see the gleam
of white houses around the port of Bastia almost forty miles away.
It was also very remote, for it could only be reached by a rough
path which climbed steeply out of Marciana Alto and brought the
pilgrim to the verge of a precipice. There was nowhere else on Elba
which offered such privacy.

While Napoleon was planning this retreat he kept dropping
hints that his wife and son might even arrive while he was in the
mountains, and that he certainly hoped before long to see them
installed in the new apartments at the Villa Mulini. On 9 August, indeed, he sent a Polish colonel named Téodor Lacynski to Aix-les-
Bains, ostensibly to deliver a letter to Marie-Louise. At the same
time he told Bertrand to write to Méneval to say 'that I expect the
Empress at the end of August; that I wish her to bring my son; and
that it is very strange that I haven't any news. This ridiculous interference is probably the work of some subordinate official, and
doesn't come from her father'.

All this talk of an early family reunion was actually nothing but
deceit, and the rough-and-ready arrangements for the camp were
proof of it. Napoleon had told Bertrand to spend a few hundred

francs to put doors and windows on the derelict building, and to freshen it up, but if he had really been expecting his wife and child he would not have planned to receive them half-way up a mountain, with only a few trusted servants, and his mother installed in a house on the edge of Marciana Alto to give an air of respectability to the whole bizarre adventure. The visit to the hermitage was actually a cover for a very different piece of family business. Colonel Lacynski had not gone to Aix to fetch the Empress; he was returning with his sister, Maria Walewska, who had long been known as the Emperor's 'Polish wife' and the mother of his illegitimate son Alexander Walewski, and Napoleon was doing all he could to keep her visit a secret.

Napoleon had met Maria Walewska in Warsaw in January 1807, when he was at the peak of his success and she was a romantic, patriotic girl who had been married at the age of eighteen to the rich septuagenarian Count Walewski, and had become convinced that she might persuade the great man to support her country's struggle for freedom. Napoleon flirted with her, and with the idea of Polish independence, which he treated with the same teasing cynicism that he always showed towards the cause of Italian unity. Before long she went to Paris as his recognized mistress with a house of her own near the Tuileries.

Napoleon was undoubtedly attached to her for a time, but he had little respect for women. He once said they found power 'the greatest of all aphrodisiacs', and he brought little more to a relationship than his fame, a style which was at once condescending, unctuous, and gauche, and a personality in all senses ill-equipped for the game of love. He was emotionally immature, fearful of impotence, and calculating about his affections. 'Yes, I am in love, but always subordinate to my policy', he said to his brother Joseph in the autumn of 1809, when he was considering divorce from the Empress Josephine in order to marry a woman who could give him a child, and Maria Walewska was already pregnant. 'And though I would like to crown my mistress I must look for ways to further the interest of France.' That meant a dynastic alliance with Austria, and marriage to Marie-Louise.

Maria Walewska returned to Poland after her son was born in May 1810. She saw Napoleon briefly on his way home from Moscow, and in 1814 she was back in France, hoping to see him again and even to share his exile. It was a vain hope: at the time of his abdication she apparently spent an unhappy night in an ante-room at

Fontainebleau and had been sent away without a friendly word of farewell. Yet that disappointment did not shake her loyalty, and when she had business in Naples—where Murat was threatening to confiscate the estates which Napoleon had settled on his natural son Alexander—she decided to call at Elba and repeat her offer to join the Emperor. It was, in fact, a letter she sent from Florence which set Napoleon planning the hermitage meeting on 27 July, and when Téodor Lacynski left Portoferraio on 9 August with the letters to Marie-Louise he was also carrying Napoleon's reply to Maria Walewska. 'I will see you here with the same pleasure as always, either now, or on your return from Naples,' he wrote guardedly. 'I will be very glad to see the little boy, of whom I hear many good things.'

★

Napoleon did not go to the hermitage until the last week of August, and even then he had to wait several days, keeping himself amused by a little hunting and by the usual flow of instructions to Bertrand and Drouot. Some were about business—a contract for the sale of iron ore in Liguria; some were domestic—exact details of the building work at the Mulini, preparations for an apartment for him in the fortress at Longone, where he proposed to stay for the town's feast-day on 9 September, the valuation of the vineyards at San Martino; and there were military matters as well—unrest among the troops and a run of desertion among the Corsicans which made him threaten to shoot the ringleaders.

But at five o'clock in the afternoon of 1 September, when Napoleon sighted a brig making for the Portoferraio roadstead, the waiting was over. The brig had been told to keep away from the port and to set the passengers ashore discreetly at San Giovanni, where Bertrand was to meet them with a carriage, but the unusual anchorage attracted rather than distracted attention, and people were soon saying that the brig had brought Marie-Louise to Elba at last. Bertrand, indeed, was so busy trying to check this unwelcome speculation that he was delayed, and it was well after nine when he lit a fire on the beach as a signal for a boat to bring in Colonel Lacynski, his sister Emilia, his sister Maria, and her small son Alexander. Within the hour the gossips had turned the four visitors into Prince Eugene, a lady-in-waiting, the Empress, and the King of Rome.

The travellers made their way to Marciana by the faint light of a quarter-moon, and Napoleon was waiting with a set of horses to carry them up the steep track to the hermitage. It was a strange trysting place, and a strange occasion. One account described Maria, well-dressed and bejewelled, taking a late supper alone with the Emperor. Another claimed that Fanny Bertrand was there as the hostess, and that she and Colonel Lacynski went down to Marciana for the night, leaving the two sisters in the roughly furnished cells of the hermitage, and Napoleon and Bertrand sleeping nearby in tents. Had Napoleon brought his former mistress to the hermitage for a romantic assignation? His valet, for instance, was quite explicit. 'It is surprising that he, who had such a liking for secrecy, should have behaved in what seemed to me a rather incautious manner,' St Denis wrote much later. 'Both evenings of Maria Walewska's visit he came out of his tent wearing a dressing-gown and went to her room where he stayed until daybreak. It was obvious to all what was going on ... Even the most humble of his subjects would have been more skilful at conducting a secret love affair than the Emperor.'

Though St Denis and other talebearers were certainly writing with an eye to readers who expected to find such revelations in their memoirs, Maria may well have gone to Elba with hopes of a reconciliation. She was still attached to Napoleon, she had been at Aix, and would have known that there was little chance of Marie-Louise joining him, and she was free, for her husband had died in June. She certainly spent a sentimental afternoon at the camp, wandering through the chestnut groves with Napoleon while he played affably with her little boy; and in the evening the Polish officers who had accompanied him to Elba came up to visit the Lacynskis and to entertain the company with national songs and dances. But she had also gone to see Napoleon on definite business, which concerned her son's property, and possibly to carry clandestine messages as well. Like Pauline she was a trustworthy courier, who had come through France and Italy and was going on to see Murat. If she was telling the truth afterwards, and not simply seeming wise after the event, Napoleon gave her a rare hint of his intentions: 'he considers his exile temporary', she noted later, 'and the information he demands is what he needs to choose the most propitious moment to bring it to an end'.

The visit must have had some value for Napoleon, or he would not have sanctioned it; and he was no doubt touched by Maria's loyalty, though his behaviour suggests that he was less the suitor than the sought. Yet he was most anxious to avoid being compromised by her presence. He certainly did not want to become the subject of scandalous talk on Elba, where he had so far shown such a regard for the proprieties that officers who lived with their mistresses were not received at the Villa Mulini. He did not wish to become the laughing-stock of the courts of Europe. He found it bad enough to be forcibly separated from his wife without being mocked as a false Emperor consorting with a false Empress in the tiniest of kingdoms. Nor had he quite abandoned his hopes that Marie-Louise might somehow manage to join him, and that some turn of events might enable them to make yet another alliance with her father. Yet he had put all these things at risk by this furtive meeting, and on the morning of the second day he was upset by a long letter from Drouot reporting the gossip in Portoferraio, and by the arrival of the mayor of Marciana, who had climbed up to pay his respects to the presumed Empress. Napoleon was embarrassed and angry, and he made it clear that Maria must leave that night.

He put a good face on things at dinner, but he was clearly eager for the Lacynskis to leave before the rising wind prevented the brig from taking them off at Marciana. The visit was over, so far as he was concerned, and he only went half-way down to the port before he said goodbye. When the party reached Marciana, however, a *burrasca* was blowing from the west, it was too rough to row out to the anchorage, and the brig was sent round to the more sheltered water at Longone; Maria, her son, and her sister then set out on a coach journey through the darkness, over seventeen miles of bad track, lit only by the flare of the torches carried by the outriders. When they reached Longone conditions were still so bad that it was only when Maria insisted that a fishing-boat put out to set them all on board the vessel. She had come and gone like a revenant from Napoleon's past.

★

Napoleon's attempts to discover what had happened to his wife, and what she proposed to do, had been frustrated all summer by normal delays in the post, by his uncertainty about her address, and by the interference of the Austrian police, who held up and even confiscated the letters they sent to each other. The only reliable line of communication was the correspondence between Bertrand and Méneval, but that was more formal than informative because both parties knew that every word would be seen by the censors. It was early July before three letters from Marie-Louise reached Elba, and four others sent from Vienna never arrived at all. It was clear that she still intended to go to Aix, and when Bertrand wrote on 3 July to protest that the waters at the Tuscan spas were just as good and the towns were much closer to Elba she had already been four days on the road, travelling under the incognito of the Duchess of Colorno. Her father, it seemed, was glad to have her out of the way while the princes and potentates were gathering in his capital to dispose of her husband's empire, provided that she left her son for surety; and she was equally glad to have her liberty again, after being kept under close watch for three months.

She travelled in a leisurely way, making a point of visiting Napoleon's relatives—Prince .Eugene at Munich, Jerome at Payerne, Joseph at Prangins on Lake Geneva—and she did not reach Aix until 17 July. Two staging-posts from the resort she was met by Count Adam von Neipperg, a handsome one-eyed Austrian some twenty years older than herself, who had a great reputation as a soldier, a diplomat, and as a man with a most winning way with women. He had lately married a mistress who had previously borne him five children. A fine talker, charming or passionate as occasion demanded, a good dancer and pianist, most intelligent and seemingly devoid of scruples, he had been chosen by Metternich to serve as Marie-Louise's personal attendant on terms which gave him a great deal of discretion. 'With all necessary tact', Metternich wrote in his instructions, 'the Count von Neipperg must turn the Duchess of Colorno away from all ideas of a journey to Elba, a journey which would greatly upset the paternal feelings of His Majesty, who cherishes the most tender wishes for the well-being of his well-loved daughter. He must not fail, therefore, to try by any means whatsoever to dissuade her from such a project . . . and, if the worst comes to the worst and all his efforts prove vain, he will follow the Duchess to the island of Elba.'

Neipperg put himself out to be pleasant and helpful, and Méneval's departure to visit his pregnant wife in Paris made it all the easier for this accomplished courtier to take charge of Marie-Louise's household. He had, in fact, been given a role as both spy and custodian which was even more ambiguous than that which Campbell was playing in Elba, for the phrase 'any means whatsoever' was clearly written with his physical attractions in mind. The unctuous Francis was apparently prepared to connive at his daughter's seduction if that was the only way to keep her from her undesirable husband. But Marie-Louise was at first unaware of

Neipperg's purpose. 'I am very satisfied with General Neipperg whom my father has put close to me,' she wrote disingenuously to Napoleon. 'He speaks of you in an agreeable manner, such as my heart could desire, for I have need to talk of you during this cruel separation; when can I at last see you, embrace you? I very much desire it.'

Marie-Louise was still hoping that she could go on to Italy in September, for on 4 August she told Menéval 'I continue to expect an answer from my father to know the date of my departure for Parma', and five days later she complained that she was 'still in the most painful uncertainty' about her future. If Francis refused to allow her to go to Parma, she wrote, unaware that the French were making great difficulties about the displacement of the Bourbon claimant to the Parma duchies in favour of Bonaparte's consort, '*I shall never* consent to return to Vienna, before the departure of the sovereigns, and I shall try to get my son back with me'. She was, she then told Menéval, 'in a very critical and an unhappy condition . . . my heart is so troubled that I think the best I could do would be to die'. On 15 August, on the morning of Napoleon's birthday, she reported to Menéval that she was 'in one of my sad moods. . . . How can I be happy . . . when I am obliged to pass this feast-day, so solemn to me, so far from the two persons who are dearest to me'. Later that same day, indeed, when she had received a message from Metternich saying that she was to start for Vienna at the beginning of September, she declared that she was 'very unhappy at the idea of being forced to return . . . all the more as no good reason for it has been given to me'.

There were in fact two good reasons why Metternich should want to remove Marie-Louise from Aix as the season came to an end and her departure would seem part of the normal course of events. The first was the growing anxiety of the French authorities about her presence on the edge of France, where she certainly provided a rallying-point for Bonapartist sentiment and possibly for a plot. She rode in a carriage which carried the imperial coat of arms; her servants wore imperial livery; her visitors included the famous actor Talma, the painter Isabey, and others who had been part of the imperial circle in Paris, as well as officers from Grenoble and several more army garrisons in the frontier region. 'I just can't believe that it is solely for the sake of her health that she has come to France', a police spy reported to Paris, noting that she neither drank nor bathed in the waters. 'If one is in poor health one doesn't spend the

day walking about, going out on carriage-rides or on horseback.'
And the local prefect passed on a great deal of gossip from the villa—
that most of the household were sympathetic to Napoleon, that
the servants expected him to return, that there were secret couriers
coming from Vienna to plan some Austrian action against France,
and even that Marie-Louise was herself party to some kind of plot.
Talleyrand, certainly, was so anxious to see Marie-Louise put out
of the way of such temptations that on 9 August he wrote a note to
Metternich, concluding that 'it would suit us both if her stay at Aix
was not prolonged'.

The proximity to France was one risk. The other was that Marie-
Louise might choose to go to Italy, or even be induced to go. Neipperg's instructions certainly covered that possibility, and it was becoming clear to Napoleon that unless his wife soon left for Italy she
was unlikely ever to go. On 18 August he wrote another letter,
urging her to be on her way. 'Your rooms are ready and I expect
you in September for the grape-picking. No one has the right to
oppose your coming. I have written to you about that. Come then.
I wait for you with impatience.'

When this letter arrived, Neipperg reported confidentially to
Metternich, 'it made a profound impression on the Empress and
immediately affected her health': it seemed, indeed, that his agreeable attentions and Metternich's insistence that a visit to Italy was
'out of the question' were beginning to have an effect. There may
have been some truth in Neipperg's final comment. 'The idea of
the journey', he wrote to Metternich,' seems to inspire more fear
than a desire to be reconciled with her husband.'

*

Napoleon, of course, was not aware of this change in his wife's attitude; in any case it suited him to make out that it was only the
'vile behaviour' of her father which prevented her from joining
him. He was an opportunist who always invested in grievances, and
by the end of the summer he was clearly preparing a list of them
which might come in useful on some future occasion. On 20 August
he sent an officer to Genoa on the *Inconstant*, ostensibly to buy
furniture for his mother and uniforms for his troops, but also 'to
seize every occasion' to write to Méneval (who had returned to Aix)
and to use at least four different routes to get the letters through
the screen of censors. At the same time Napoleon gave a month's
leave to Captain Hurault de Sorbée, whose Austrian wife was

Marie-Louise's *lectrice* and could thus introduce him to her directly. This was a serious attempt to make contact, for Hurault was specially briefed on means to avoid surveillance and given a secret letter in which Napoleon told Marie-Louise that the *Inconstant* would be waiting for her at Genoa on 10 September; if she agreed to escape from Aix then Hurault would make the necessary arrangements.

Hurault travelled fast, reaching Aix on the evening of 26 August, and handed over the lettter. But Marie-Louise had now found cause to change her mind, for Neipperg had become her intimate confidant; when she showed him the letter Hurault was threatened with arrest for intent to abduct and despatched to Paris for interrogation. On the same evening that Maria Walewska arrived in Elba the Empress wrote a compliant letter to her father. 'Be assured', she said, 'that I am now less than ever desirous of undertaking that voyage, and I give you my word of honour that I will never undertake it without first asking for your permission.' A week later she was even more forthright, telling the Duchess of Montebello that the idea of 'an escapade' to Elba was ridiculous. Napoleon's repeated requests for her to join him, she wrote, had 'no influence on the Court in Vienna, except to make them keep me away still longer from Parma. I shall give the ministers my most sacred word of honour that I shall not go to the island of Elba at the moment, that I shall *never* go (for you know better than anyone else that I have no desire to do so)'.

Neipperg had clearly done what he had been sent to do, for Marie-Louise was in a cheerfully complaisant mood when they left Aix together on 5 September, setting off for a tour of Switzerland which was soon to become more like a honeymoon than a dutiful return to Vienna. By the time they reached the Bernese Oberland, discarding their attendants as they went, they were clearly on very close terms; and on the night of 25 September, when they were staying at the Soleil d'Or at the foot of the Rigi, Marie-Louise sent the customary footman away from the door of her bedroom. Soon after she reached Vienna her relationship with her escort was so apparent that the gossips openly called her 'Madame Neipperg', and Neipperg himself was formally appointed as her chamberlain to provide a public reason for his constant presence at her side.

Napoleon must have known for some time that he could no longer hope to see his wife and child again, at least in Elba; but he still kept up the appearance of hope. When Campbell had a long audience with him at Longone on 16 September he asked whether there was any news of them, speaking 'with warmth, and in strong language, upon the interdiction to their joining him', calling this an 'instance of barbarity and injustice' without a modern parallel and insisting that Marie-Louise was being held against her wishes. 'She was now absolutely a prisoner', Campbell reported Napoleon as saying, 'for there was an Austrian officer who accompanied her, in order to prevent her escaping to Elba', and he pointedly asked Campbell whether Lord Castlereagh approved of these constraints.

In the event, however, neither Campbell nor Castlereagh nor any other intermediary could save Napoleon's marriage or his dynasty. Except for the circumscribed contact with Méneval, indeed, he could not even maintain contact with his wife once she returned to Vienna. On 10 October, in a final gesture, he sent a begging letter to Marie-Louise's uncle, the Grand Duke of Tuscany, who had recently been restored to the domains so recently ruled by Napoleon's sister Elisa. Saying that he had heard nothing from his wife for two months, or from his son for almost six months, he asked 'whether you will allow me to address to you every week a letter for the Empress, and will send me news of her in return?' The Duke did not even acknowledge this letter from the man who regally addressed him as 'My Brother and Very Dear Uncle', but he sent it on to the Emperor Francis. Francis merely showed it to his daughter, with instructions that she was not to reply. And she never did. She never wrote Napoleon another word.

11

A MAN OF ORDINARY TALENTS?

'On a small island, once you've set the machinery of civilization going', Napoleon said plaintively some years after he left Elba, 'there's nothing left but to die of sheer boredom—or to get away from it by some heroic venture.' The phrase was a neat summary of his attitude as 1814 slipped away. During his first months of exile he had behaved more like a rich man taking a holiday in his newly acquired estate than a defeated ruler thrust into humiliating retirement. As late as 13 September he told Bertrand to acquire the peninsula on the south coast called Capo Stella, and to fence it off as a hunting reserve. But this was almost the last of such schemes of improvement, for the cramping realities of life in his petty kingdom were closing in on him. Conscious of his unused talents, of his compulsion to work (Pons said that he toiled away 'as though time not completely full was truly time lost'), of his incomparable knowledge of politics and his unique experience of war, he was cooped up like a garrison commander in a rough and poor-mannered place where he squandered his energy on trivialities.

He was generally up before dawn, going over despatches, dictating notes or orders, or looking at newspapers and intercepted letters. Like his father-in-law the Emperor Francis, he had an appetite for police reports, and he was always ready to put other things aside to listen to one of his informers, even if the man brought nothing but kitchen or barrack-room gossip. 'He liked every bit of vulgar chit-chat,' Pons recalled, 'all kinds of foolishness.' After breakfast at seven he would take a short nap before riding off to inspect his troops, or look at storehouses and fortifications, or watch the men working on the house and grounds at San Martino. On his return he often held an audience, especially if there were visitors from the mainland, took a light lunch with Drouot, and retired for a short siesta or a read. In the afternoon, before dining promptly at six, he drove out into the countryside or had himself rowed about the harbour in the barge that Captain Ussher had given him as a parting gift.

All this activity helped to break the tedium of days without purpose or consequence, but the autumn evenings were generally very dull for there was no one in the royal household, or among the small group of Elban officials who were expected to dine regularly at the Villa Mulini, who had the vivacity to enliven them. Napoleon himself was often moody at the end of the day. On most nights he would sit quietly for an hour or two, playing cards or dominoes with his mother, taking on Drouot for a game of chess, or talking in a desultory way to one of the court chamberlains, and about nine he would give his customary signal that he was about to retire by picking out the first notes of Haydn's 'Surprise' symphony on the piano. It was a dreary daily round for a man who had lived at such a pitch of excitement, with all Europe for his stage and kings hanging on his words, but he had long ago learned the value of dissembling patience when he was unsure what to do next. He would have to bear the boredom of his exile until the Allies changed their minds or he could change his fortunes.

He was not a man to discuss such feelings, even with intimates, and none of the men around him had any idea whether they were going to stay on Elba or whether they were merely going through the motions of settling down while the Emperor waited to see what happened. The three generals, for instance, had very different views of their prospects. Drouot had hoped to retire to a life of reading and reflection, and until he learned that his mother in France disapproved of the match he had hoped to marry the young and attractive Henriette Vantini, who had been teaching him Italian. Cambronne, on the contrary, found Elba a frustrating place. He was most at home on the battlefield, and he shared the hope of the Guard for anything that would take them back to it.

Bertrand was more ambivalent. He had no liking for the wearying and tedious life he led on Elba. Since he was grand marshal—a post which in these modest conditions combined responsibility for the royal household with the roles of prime minister and chief of staff—Napoleon expected him to be constantly at hand, ready to deal with visitors, control the staff, provide information, take dictation, or put an order in train. His work was never done, and it was always difficult for him to get away from the Villa Mulini and back to his family, who were living in the ill-furnished apartment in the town hall where Napoleon had first stayed. His wife Fanny, moreover, was wretchedly depressed after a tiresome journey across France and the death of her baby soon after she arrived in Portoferraio; she was reluctant to play any part in the little court at the top of the hill, and she was convinced that Bertrand had made a serious mistake in accompanying his master to Elba. Both the Bertrands, therefore, had good reasons to see the island as no more than a temporary retreat. Yet they both feared that any change might be for the worse.

★

Life had become tedious for Napoleon and his staff, but it had become much livelier for his subjects, for there was always some novelty to distract them; and the presence of the Emperor brought people flocking to Elba in such numbers that the *Inconstant* sometimes came in crowded to the bulwarks, and even the most respectable visitors had to make do with a bunk or a shakedown in one of the public rooms of the Auberge Bouroux. There were Corsicans who came to beg favours from Letizia Bonaparte, half-pay officers who came seeking commissions, a Frenchman who claimed that he wanted to start a newspaper, and a man who proposed to erect ready-made huts on Pianosa, merchants, mendicants, and old admirers of Napoleon who simply wished to pay their respects to him.

There were some very odd women as well as men among the newcomers who were scrutinized by Cambronne and the gendarmerie. What were they to make of the dubious Comtesse de Polignac and the genuine Countess of Jersey, of the mysterious Madame Theologos, a handsome Greek who became the mistress of Peyrusse while her husband—possibly a secret agent—was busy compiling a dictionary at Napoleon's expense? And what were they to report about Signora Filippi from Lucca, who had dressed as a man and served in the army of Italy before following the Emperor to Elba; of Madame Giroux, a stout-hearted patriot of sixty who had travelled all the way from Versailles to show her sympathy for her fallen idol; about Madame Darcy, an attractive woman of twenty-five who made a set at Napoleon; and about the oddest of them all, the Comtesse de Rohan-Mignac?

This plump adventuress in her forties had been a brothel-keeper in Paris, and she had acquired her title by marrying an aristocrat whom she had hidden during the Terror. She arrived from Malta with her twelve-year-old son and set up in some style, taking a set of rooms, driving about in her own landau, calling at the Villa Mulini, on Napoleon's mother, Madame Bertrand, Madame Pons, and other ladies in Portoferraio's very small society, and attracting officers of the Guard to her own receptions. Napoleon thought little of her: 'only *gnic-gnac* there', he said disparagingly. All the same she managed to get herself invited to the reception he gave to celebrate Pauline's return from Naples, and she secured a seat quite close to him, bringing her child to the table in such a way, Pons felt, as to imply that he was the Emperor's natural son. 'She presumed too much, and drank too much', Pons said. As she left Elba, a rumour that she had been exposed as a double agent went with her to Livorno.

★

The English travellers who began to turn up in Elba in the summer of 1814 were very different from this raffish company. Most of them were men of rank and influence—Whig noblemen, politicians, officers on leave or well-to-do gentlemen making the grand tour who hoped to include a sight of Bonaparte among other curiosities; and Napoleon welcomed them all. It was actually easier for some Englishmen to speak to their former enemy than to their compatriots, it seems, for Pons complained of the time he had to waste showing snobbish Englishmen over the iron mines, and he noted with the sardonic eye of an old Jacobin that they did not speak to each other unless they had been formally introduced. Napoleon, by contrast, made himself agreeably accessible. Even if a visitor was not sufficiently important to warrant a formal audience or a dinner at the Villa Mulini he was always willing to chat casually by the roadside when he was out riding.

'Can this be the great Napoleon?', a young tourist from Suffolk named J.B. Scott asked himself, when the King of Elba stopped to talk to him and four British officers near Longone on 19 September. 'Is that graceless figure—so clumsy, so awkward—the figure that awed emperors and kings? It is surely impossible; and that countenance; it is totally devoid of expression; it appears even to indicate stupidity! Such was the *first* impression, and though I soon found reason to change my opinion concerning his countenance, I still continue to think the figure of Napoleon very unmartial.' Like many people who saw him at this time Scott thought him very paunchy, so addicted to snuff that his clothes were stained with it, and very seedy-looking—a doughy complexion, with his hair hanging 'very long in candle-ends', a cocked hat whose 'brownness seemed to indicate that it had stood many a campaign', tarnished decorations, an old coat, shabby boots, and 'a dirty bridle and bit' on his horse. Yet as Napoleon began to speak 'quickly and incessantly', asking each officer in turn about his military service, Scott began to change his opinion, deciding that 'his eye and voice inspire respect, and his manner indicated great talent ... his smile gives confidence to those who hear him'.

This forthright way of talking surprised people who expected Napoleon to be reserved and condescending, and it was so flattering that they found themselves forgiving his blunders and forgetting his crimes. Captain Ussher had been one of the first Englishmen to succumb to this magical charm, but other naval officers soon proved so susceptible that Campbell had them forbidden the island and a French royalist dismissed them as sycophantic imbeciles. 'If you had called up the ghost of Caesar or Hannibal', an English visitor remarked after a long interview in November, 'you should have expected to have heard them talk in the same style'; and he added that the Emperor was so easygoing that he felt that he had been talking 'to an ordinary person'. The effect was heightened by apparent frankness. 'He answered on all subjects without the slightest hesitation, and with a quickness of comprehension and clearness of expression beyond that I ever saw in any other man,' Lord Ebrington concluded; 'nor did he, in the whole course of the conversation, betray, either by his countenance or manner, a single emotion of resentment or regret.' Lord John Russell, the future Whig Prime Minister, went to the 'wretched palace' in Portoferraio for an audience a week after Ebrington had dined there, and he came away with the same impression. Napoleon, he said, was 'very fat, without much majesty in his air and still less terror in his look', good-natured, and so open in his manner that 'during the two hours I was alone with him [he] talked and encouraged me to talk on every subject'.

Napoleon was far from being as frank as Ebrington and Russell believed. He was actually a most calculating man, who had the knack of being garrulous without being indiscreet, and these seemingly artless conversations, like his long discussions with Campbell on 3 August, 16 September, 31 October, 4 and 21 December, served several useful purposes.

It obviously suited Napoleon to have a succession of well-connected Englishmen visiting Elba. Lacking any kind of diplomatic recognition except Campbell's ambiguous appointment, he thought their attentions improved his standing generally and particularly in England. From the moment when he had asked Castlereagh for asylum in England, in fact, he had clung to the bizarre notion that the English might be willing to offer him a more congenial home than Elba. In England, he hopefully told Campbell in October, 'he would have society, and an opportunity of explaining the circumstances of his life, and doing away with many prejudices', and when he received Lord Ebrington he specifically asked what would happen to him if he went to London, and whether the mob would stone him. 'I replied', Ebrington wrote, 'that he would be perfectly safe there, as the violent feelings which had been excited against him were daily subsiding, now that we are no longer at war.'

Napoleon also wanted the English to believe in his peaceful intentions. 'I want to dispel any illusions. I think of nothing outside my little island', he said to Campbell on 16 September, in much the same terms as he spoke to other Englishmen who saw him in the course of the autumn. 'I no longer exist for the world. I am a dead man. I only occupy myself with my family and my retreat, my cows and mules.' This mock modesty was so persuasive that it even had some effect on the sceptical Campbell. 'I begin to think he is quite resigned to his retreat, and that he is tolerably happy', he wrote on 20 September; and other Englishmen were even more credulous. The war had left them so politically divided that the Whigs almost preferred Bonaparte to the Prince Regent, and the parliamentary reports and opposition newspapers in London were full of complaints that he had been treated too harshly, and that the victors rather than the vanquished were to blame for the troubled state of Europe. On 20 August, for instance, William Cobbett capped a series of sympathetic comments in his *Political Register* with the claim that Napoleon was doing the same things in Portoferraio that he had done in Paris—beautifying the town, drawing up a constitution, drafting sound laws, encouraging the arts and sciences, and disbursing 'large sums of money on useful undertakings, instead of spending it on pimps and parasites'.

Napoleon certainly hoped that reports from his visitors would help to allay the fears that he was likely to burst out of Elba like a demonic jack-in-the-box; but he was also eager to elicit information from them which would be useful if and when he decided to quit Elba. That was his direct purpose in talking to his English visitors. He carefully quizzed them about the political situation in England, or the impression they had formed as they travelled through France and Italy, and when he expressed his own opinions he watched closely to see how they reacted.

★

The long discussion with Campbell on 16 September was a good example of this technique. 'He constantly walked from one end of the room to the other', Campbell noted, 'asking questions without number, and descanted upon a great variety of subjects.' Since Campbell had just come across from the mainland Napoleon naturally began by asking about conditions in Italy, and Allied policy towards Murat, Spain, and the future of Corfu. He wanted to know whether Campbell had news of Marie-Louise, why the Austrians were so afraid of his modest attempts to recruit more soldiers in Italy and Corsica when his troops 'were not sufficiently numerous to guard all the villages and fortifications', and what was happening in France. Campbell replied that the Bourbons had done well in some respects, though the men who had been turned out of their posts by the new government, like the returned prisoners and the lower ranks of the army, 'were attached to him'. That, of course, was a very delicate topic, but Napoleon always dealt with it in much the same way and ended with much the same warning. Louis XVIII, he claimed on this occasion, would be ruined if he became 'the English Viceroy' and let the Allies rob France of Belgium and a natural frontier on the Rhine, for the French people would not indefinitely tolerate such humiliation. 'Their chief failings were pride and love of glory', he told Campbell, 'and it was impossible for them to look forward with satisfaction and feelings of tranquillity, as was stated to be the sincere wish of all the Allies, under such circumstances.' In the middle of November, when he saw Mr Fazakerley and his friend Mr G.W. Vernon, a Whig MP who was a cousin of Lord Holland, he was even more explicit about the prospects of trouble in France. After a four-hour talk 'about emperors, and kings, and revolutions, the loss and acquisition of kingdoms', he came to the realities of his own country. 'You will see,' he said. 'These days a wind of liberty is coming out of the villages, which will overturn everything.' It was an ominous remark, and its effect was scarcely lessened by the way he shrugged off responsibility for anything that might have happened. 'That doesn't concern me,' he told Fazakerley and Vernon. 'I am like a dead man; my role is finished.'

Despite such disclaimers Napoleon had a lively and well-informed interest in everything that was happening on the Continent. In his interview with Campbell on 31 October he began by talking about the strength of Italian nationalism and criticizing the malign influence of 'the Pope and his priestcraft'. Then he turned to more serious matters. The Congress had started in Vienna. How long did Campbell expect it to last? His own answer to that important question, on which so many matters of timing hinged, was that it would be fairly brief, and that it would do little more than ratify agreements that the Allied sovereigns had already reached among themselves in private. It would have 'a dreadful effect' on Poland, Italy, and France if the victors were to separate 'without any final settlement of Europe'.

Thus far he had spoken more like an elder statesman than an interested party on two of the three subjects that were so plainly linked in his mind—Italy, the capacity of the Allies to settle issues which unresolved would soon lead to a new war, and the position of France. Once again it was not so much the capacity of the Bourbons to govern which seemed to interest him as the theme which he persistently regarded as the most sensitive of all: he spoke angrily to Campbell about 'the state of humiliation to which France is now reduced by the cessions and the aggrandisements so unequal, of the other leading powers!' The appointment of the Duke of Wellington as British ambassador in Paris was 'another insult and injury to the feelings of the French people' but the real cause of their 'universal disgust' was the Belgian frontier, which had been the sticking point for him in all the peace negotiations in the last winter of the war. France needed the line of fortresses to cover Paris. She needed Antwerp even more, for the port's capacity for trade and its capacity for building a fleet that could rival the Royal Navy were vital to any future confrontation with England. There was thus no prospect of 'a lengthened state of tranquillity' if Antwerp were handed to Holland to satisfy England's demand for security.

In this fascinating conversation Napoleon actually revealed the way his mind was running in the autumn of 1814—keep the Italian nationalists in play as a distraction; wait to see how the Allies agree or disagree at Vienna; let discontent with the Bourbons simmer in the barracks and villages of France; and stand firm on Belgium, so that any challenge to the Bourbons could be made in the name of glory and the natural frontiers. There was no definite plan behind it, no hint of secret dealings, and nothing to suggest an intention to strike at a specific time or place. Yet it was a coherent programme, and Campbell failed to see the logic of it.

One reason for this failure was Campbell's conviction that Napoleon was far more likely to cause trouble in Italy than in France. Another was his professional incompetence. While he was an earnest reporter he lacked the capacity to evaluate what he heard and saw, and the more bored he became with his life on the island the more he was prone to ignore signs of danger and to persuade himself that he could safely go away and neglect his duties. The most important reason, however, was a defect in Campbell's character. While most people were bamboozled by Napoleon's calculating charm, he let dislike colour his judgement. 'Campbell thinks Bonaparte a man of ordinary talents who has had a great deal of luck', J.B. Scott noted during his visit to Elba in September; and jejune remarks of this kind are scattered all through Campbell's journal and his reports to Castlereagh. Whatever else explained Napoleon's extraordinary success it was neither luck nor ordinary talents that had made him master of Europe before he was forty.

★

There was much talk on Elba about a special English interest in its king. After all, he had arrived in an English frigate; Colonel Campbell, with another English warship at his disposal and direct access to the Villa Mulini, was the only Allied representative on the island; there was always a welcome for English visitors—even when there was a sudden tightening of security in September, and other arrivals were turned away, the English were admitted; and on several occasions one of Napoleon's adjutants noted the hearsay that a messenger had arrived from England with secret despatches. The English were so favoured, indeed, that people said they must have struck a bargain with their old enemy and were protecting him for some mysterious purpose. 'It is universally supposed in Italy, and publicly stated', Campbell wrote plaintively at the end of December, 'that Great Britain is responsible to the other Powers for the detention of Napoleon's person, and that I am the executive agent for this purpose.'

Both these notions were dangerously misleading, and Castlereagh was responsible for the confusion. In April he had formally refused to take any responsibility for the fate of the man he recognized only as General Bonaparte. He had declined to sign the treaty that made Napoleon the King of Elba, explicitly rejected the idea that Campbell was in any sense his gaoler, and made it clear that Britain was neither willing nor able to impose a peacetime blockade on an island which had been properly ceded to its new ruler. At the same time Castlereagh had taken responsibility for all the informal arrangements, without providing any clear instructions or proper support for them, and apparently without more than a casual thought for their implications.

What had begun as an error, committed under the press of business as the war ended, thus became a more and more serious mistake as the months passed, and by the end of the year it had developed into the most puzzling aspect of Allied policy towards Napoleon. The British had no legal rights in the matter, as long as he refrained from making mischief. It was even doubtful whether an English captain could properly do anything if he actually found Napoleon at sea in the *Inconstant*. Yet their improvised involvement on Elba had lulled them and everyone else into a false sense of security. As long as Campbell remained on or near the island, with a sloop to hand and other naval units on call at Palermo and Genoa, it seemed that the delegates in Vienna could take their time resolving their differences.

Perhaps Castlereagh simply assumed that before the fallen emperor could rise again Campbell would be able to sound an alarm.

12

THE GAME OF SECRETS

News from Elba travelled so slowly, and was so unreliable, that most of the information which reached the Allies in the course of 1814 was late, trivial, misleading, contradictory, or plainly false. While Campbell had his failings—he was away too much, he was obtuse, and he was an amateur playing a professional at the game of secrets—he at least sent Castlereagh dull but reasonable accounts of what Napoleon was saying, and what he appeared to be doing. Almost all the other agents who were set to watch the King of Elba were rumour-mongers, or worse, retailing any kind of rubbish that came their way. It was variously reported that Campbell had seen Princess Pauline wandering about Livorno dressed as a sailor, that a one-eyed Jew had been sent to assassinate the King of Elba, that General Köller, Prince Eugene, Lord William Bentinck, and even Marie-Louise had made clandestine visits to the island, and that Napoleon hoped to escape to the United States with the help of a Barbary corsair. It is not surprising that the credulous soon came to believe everything that was reported from Elba; the sceptical believed nothing at all.

The failure to establish a coherent system of surveillance over Napoleon was a most serious yet understandable blunder, for the Allies were in such a muddle about him, and so overextended by the final collapse of his empire, that they were quite incapable of common action. There were agents from different countries spying on the Emperor, and on each other, there were half-baked plots to abduct Napoleon, political conspiracies that were largely composed of provocators and dupes, and ample opportunities for adventurers, liars, and imposters to try their luck.

Nothing was done properly. Sovereigns who were ready to spend a small fortune on dinners and dances, on presents to their mistresses and bribes to foreign statesmen, were misers when it came to paying their spies. One of the most effective French agents complained at the end of the year that he had not even been paid his out-of-pocket expenses. Few of the men responsible knew how to recruit reliable agents, supervise their work, or evaluate their reports. No doubt some of these agents now worked as zealously against Napoleon as they had served him in the past, but in the confused circumstances of 1814 nobody could tell for sure whether a man carried a tricolour cockade in his pocket or still heard the 'Marseillaise' ringing in his ears.

Even if the Allies had deployed ample money and good men, moreover, they could not have overcome the practical problems set by Napoleon's presence on a small island so close to the mainland. It was hard enough to penetrate his defences, for he had only a few beaches to watch, he lived in a walled fortress, and one man in five on his island was a soldier or a gendarme under his orders. 'We are dealing with an enemy who knows all the tricks of the trade', the head of the French police declared on 18 August, 'and one cannot hope to cope with him by routine police measures: he has employed them himself too long to be taken easily.' It was even harder to stop him communicating with anyone he wished, for the long and poorly patrolled coastlines of France and Italy were open to him, and there were so many ways of making contact across the short sea-crossings that he had no need to trust confidential messages to the post. Jacques-Claude Beugnot, the royalist director-general of police, had realized as early as 5 July that it was useless to search all the correspondence with Elba. 'Experience has shown a hundred times', he wrote, 'that even under the inquisitorial despotism of Bonaparte anyone who is really determined to hide a letter can escape all the searches of the authorities.' In any case few of the Allied agents on or near Elba seems to have known exactly what their masters wanted them to find out or do about Napoleon. Like Campbell, they were generally left to their own devices, in a state of generalized suspicion, trying to make sense of fragments of information. The situation was so unsatisfactory, indeed, that even sensible questions sounded like riddles. Clandestine activity is never the best field for allies to display common sense or to develop a common policy.

*

The Austrian secret police was a well-organized bureaucracy which relied heavily upon room searches, eavesdropping, the opening of letters, the decoding of simple cyphers, and the seizure of couriers. But such methods, as Baron Hager conceded in a plaintive letter to the Emperor Francis, were much less effective against Napoleon on Elba. The Austrians learnt very little from the letters they opened, or from the reports their consuls at Livorno and Civita Vecchia made about travellers to Elba, or from occasional arrests. The recruiting officers they seized in July told them nothing and their only substantial catch was Dominico Ettori, an unfrocked monk originally sent to Elba by Marshal Bellegarde, who gave himself such an air of mystery when he returned that he was suspected of being a double agent and sent off to Vienna for interrogation. They learnt rather more, in fact, from their ambassadors—Lebezeltern at the Vatican and de Mier in Naples—who picked up scraps of news about Napoleon from his relatives. On 1 August, for instance, Lebezeltern sent Metternich a general warning that reports circulating among Napoleon's sympathizers in Italy 'indicate next January as the time for the changes which will take place in favour of their cause and their former leader'. Yet even this prescient remark may have been no more than a lucky hit among many errors.

The police of Tuscany, previously ruled by Napoleon's sister Elisa and now restored to Grand Duke Ferdinand III, the brother of the Emperor Francis, could add little to the efforts of the Austrians who had occupied their little duchy. General Spannochi Piccolomini, the governor of Livorno, had been arrested by Napoleon many years before and his old grudge may well have prompted new suspicions. 'We have a bad neighbour', he wrote to the chief of police in Florence on 25 May 1814, 'and I believe that the government can and must have reliable reports on what Napoleon does, says and thinks.' In October Lord Burghersh arrived in Florence as the British envoy to the Tuscan government, and he was soon complaining that its police at Livorno and Piombino were so bad that the number of persons 'who pass through this state to Elba, and return without being known to the government, is very great'. Although Galli, the police inspector at Livorno, did compile daily lists of people making the crossing, they were almost useless for any purpose save proof of Galli's diligence. For Galli only caught one courier—a man from Longone named Lorenzo Piachi, who was arrested on 28 June 1814—and he seems to have been equally incapable of really harassing Napoleon's agents in Livorno and of finding informers on Elba who could tell him more than waterfront gossip.

Further down the Italian coast was Civita Vecchia, the port which served Rome, directly ruled by the Pope. Generous to Napoleon's relatives who sought asylum in the Holy City, especially to his brother Lucien, his mother, and her half-brother Cardinal Fesch, the Pope was firm against Napoleon himself, for the Bonaparte name was still a rallying cry for all the anti-clericals in Italy. He was equally hostile to Murat, a fervent freemason and commander of an army which was a standing threat to the Papal States. Yet His Holiness was so chronically short of money for spies, Lebezeltern complained to Metternich, that he could do little to embarrass either the Emperor or Murat except to keep a watch on travellers between Elba and Naples who passed through his domains. Fortunately his chief administrator at Civita Vecchia was Tiberio Pacca, the nephew of his secretary of state and a vigorously wary man.

In December the younger Pacca had two successes. The first was the staging of a dockside brawl near the *Inconstant*, in the course of which one of his men made off with Lieutenant Taillade's valise. When this was opened it was found to contain a letter from Bertrand to Lucien in which he commended a Corsican named Ramolino as the trustworthy bearer of a verbal message, and also two blank sheets of paper, one signed by Bertrand and the other by Napoleon himself. Ramolino was a relative of Letizia Bonaparte, and he had been director of taxes in Corsica before he defected to the Emperor in the autumn of 1814. He soon became one of Napoleon's regular couriers. The second coup was the interception of a letter from Cardinal Fesch, telling Bertrand to read Volume 127 in a consignment of books sent to Napoleon. When the package was opened this volume contained two letters from Murat, one to Napoleon and one to Pauline, and some notes from Fesch and Lucien. There was nothing very compromising about these letters, for Murat wrote about family affairs and Fesch was simply passing on chit-chat about Murat's attitude to the Pope and a rumour that the Congress in Vienna might end by partitioning Italy. Yet this proof of a clandestine correspondence between Murat and Napoleon, despite their public denials of any contact at all, was enough to agitate all those who anyway believed that the brothers-in-law were bound to be hatching some kind of conspiracy.

*

There was one official on the Tuscan coast who was convinced that Napoleon remained a threat to the peace of Europe and that it was a folly to leave him undisturbed in his little kingdom. This was the Chevalier Mariotti, whom Talleyrand had sent to Livorno in August 1814 to reopen the French consulate as a listening post on the Tuscan sea. Mariotti was a good choice. A Corsican who had fought well in the wars, he was one of the first officers to be awarded the Legion of Honour, and he had been attached to Elisa Bonaparte as her prefect of police when she and her husband were ruling Tuscany. Antagonized by some slight, or disappointed in hopes of further promotion, Mariotti had apparently become disillusioned with the Bonapartes and towards the end of the war he had become an active royalist agent and had helped to subvert the Tuscan garrisons.

Since Mariotti spoke Italian and had excellent local contacts and experience, as well as a knowledge of the ramifications of the Bonaparte family and the manner in which Napoleon ran his secret police, he was soon able to establish his own network of agents and to collate information from other sources. He had regular contact with the French authorities in Corsica, Livorno being one of the main ports for vessels coming from that island. He kept in touch with the Comte de Bouthilliers, the prefect of the Var, and the Marquis d'Albertus, the prefect of the Bouches-du-Rhône, both of whom were concerned about French travellers crossing to Elba, and both of whom at least temporarily placed agents there. And at the end of November he began to receive almost daily messages from a new agent, usually known as the Oil Merchant, who was on the first of two long visits to Portoferraio.

The Oil Merchant must have been an experienced spy, for he knew about codes and sympathetic inks, he was cautious about passing on gossip and rumour, and he was sufficiently effective to get a good idea of what Napoleon proposed to do some time before he did it. He was also well-chosen for his task, for he seems to have been an Italian from Lucca who had served in the Army of Italy, and there were several old comrades and compatriots on Elba who vouched for him; and he arrived in good company, for his first reports show that he crossed from the mainland with Count Litta, a leading Italian nationalist who had come to seek Napoleon's support for his cause. Both these facts suggest that he was probably a man named Alessandro Forli, who landed at Portoferraio early in the morning of 30 November and registered at Cambronne's office before going off to seek lodgings. He soon found enough customers to keep up a businesslike appearance. Before long he was selling olive-oil to Letizia Bonaparte's household, to the Vantini family, and others close to the court circle, and no one appears to have suspected him. On the contrary, people talked to him in a casual gossipy fashion that enabled him to get far closer to them than was ever possible for the stiff-necked Campbell.

His reports read rather like a diary; he made the most of his friendships with officers of the garrison. One of his old associates is the 'belle amie' of a senior officer in the Guard; the wife of another is part of the royal household at the Villa Mulini, and she passes on a hint that Napoleon might be thinking of escape. 'What do you think?' he was said to have asked Drouot. 'Would it be too soon to leave the island at Carnival time?' And though this remark may have been no more than a characteristic ruse to unsettle and confuse his enemies, the Oil Merchant notes that his military drinking companions in the Buono Gusto café are in much the same mood. According to them, 'if their friends on the mainland were ready, they could quickly return His Majesty to his throne'. A hot-headed old friend from Milan says much the same when the Oil Merchant tells him that a number of mutual acquaintances in Milan are out of work. 'Things will soon change', the man flatly declares.

The Oil Merchant was a good witness because he reported what he saw without exaggeration, and without trying to fit his impressions to some preconceived idea. Three days after he arrived some friendly officers got permission to take him on a tour of the fortifications. 'They were not so much fortresses in a state of defence', he notes, 'as stores of arms and munitions'—and though a more excitable agent might have concluded from such evidence that Napoleon was planning for a break-out rather than a siege, he is careful to point out at the same time that he saw Polish lancers being trained as gunners, using the red-hot shot that would be so deadly if hostile ships tried to force their way into Portoferraio harbour. He also had a good ear for gossip without being too much impressed by it. The French officers are jealous of the Italians, he reports, and they both dislike the Corsicans. Major Colombani, whose pretty wife is one of Pauline's ladies-in-waiting, insists that 'there is a complete understanding between Napoleon and Murat', and another informant says that unemployed officers who have come to Elba seeking appointments have been told by Drouot to go and enlist in Naples, 'for those who serve its sovereign also serve the Emperor'. One day he tells Mariotti that he has been talking to an officer who is about to leave on a clandestine mission to Milan, and gives a description so that the man may be watched; and on another day the *belle amie* of Commandant Lamoretti remarks that all the officers of the Guard are talking about the day 'when we get to Milan'.

Yet the Oil Merchant is not taken in by such loose talk of a descent on Italy. 'By and large,' he notes on 6 December, 'people say that the Emperor won't show a sign of life until the Congress is over.' He reports that a former comrade is complaining that Napoleon is doing nothing to help the Italians throw off the Austrian yoke, and that unless he acts quickly 'all will be lost'; and he writes that another, whom he had known in Florence five years before, is quite convinced that Napoleon will leave for a different destination. 'As soon as that man sets foot on the mainland', his informant claims, 'you will see France in revolt—in France one blows on the fire to make it blaze.' Not one of these reports contains a particular secret, but their cumulative effect is impressive. With a quoted remark here, a shrewd observation there, the Oil Merchant catches the mood on Elba so well that he strikingly confirms Campbell's claim that Napoleon and his followers are 'dangerously restless'.

★

Napoleon was as eager as his enemies for reliable information, and in a somewhat better position to get it. Whether it concerned his personal safety or the political and military situation in Europe, he could turn to a variety of sources. There were men who had served his clandestine purposes in the past, especially in France; there were members of his family, and all their retainers, as well as sympathizers who wished to see him return to power; and besides books and newspapers there were always visitors to provide the latest news.

Confidential agents obviously played some part in this improvised system. There are glimpses of them at work in Paris and in Vienna, or making their way to Elba as couriers, and there was a good deal of covert correspondence and message-carrying in which all the Bonapartes were involved. In the middle of August one of Napoleon's valets returned from Elba with seemingly innocuous messages for Hortense and Caulaincourt, and the famous actor Talma was almost certainly some kind of go-between, for during the course of the summer he visited Marie-Louise at Aix, Joseph Bonaparte in Switzerland, and Queen Hortense at the spa of Plombières in the Vosges. Even if Murat and Caroline were very cautious, Pauline saw a good many people and heard much news while she stayed in Naples from June to October, and Murat never seems to have made things difficult for Napoleon's informants and messengers. Letizia Bonaparte was in Rome for much of the summer, and when she left to join Napoleon on Elba he could count on Cardinal Fesch and Lucien to keep him posted on events in the Eternal

City. Napoleon's sister Elisa was closely watched in Austria, where she had been detained when she went to plead her case at the Congress, but his brother Jerome seems to have kept in touch all through the year, and his brother Joseph made his home on Lake Geneva such a centre of Bonapartist intrigue that the Allies eventually required the Swiss to expel him. Hortense, at once his stepdaughter and the separated wife of his brother Louis, was at liberty in Paris and with some of her friends had formed an active Bonapartist cabal, while his stepson Prince Eugene had gone to Vienna and found himself on increasingly cordial terms with the Tsar. In the first weeks of 1815 Baron Hager often noted that the Tsar and Eugene had dined together, or walked together, or attended a ball; and on 23 January he reported to the Emperor Francis that his secret police had intercepted a message, hidden in the handle of a brush sent by Hortense to Eugene, which was said to list the names of sympathetic marshals in France and the number of troops under their control.

Marie-Louise was not involved in these rather amateurish proceedings. She had cut off all contact with Napoleon, and according to Baron Hager's spy at the palace of Schönbrunn had become so enamoured of Count Neipperg that she spent most of her time with him, but at least one of the French members of her entourage managed to keep in touch with the Emperor. 'Notwithstanding the rigorous surveillance that was kept up around me', her secretary Méneval wrote afterwards, 'I sent him news by every possible channel. I found facilities in the commercial traffic with Vienna. Kind merchants, whose hearts had not been hardened by politics, readily undertook to transmit my letters to General Bertrand. . . . All over France and Italy similar pretexts of commerce marked the correspondence with the island of Elba.'

Napoleon could also count on the family connections of his subordinates and on his friends. General Bertrand's brother, for instance, stopped briefly at Portoferraio on what passed for a holiday journey to Italy. 'Our relations with France and with our families were never interrupted', Peyrusse frankly admitted in later years; and the same was true of the soldiers on Elba. They wrote to comrades in the garrisons and villages of France, and letters came back, and men went home on leave.

★

The Englishmen who talked to Napoleon on Elba were able to

report what he said at the time, and often at length, but many of his visitors came for more confidential reasons. On 28 September, for instance, a young man named Jean-Baptiste Dumoulin, who was the son of a prosperous glovemaker in Grenoble, arrived secretly in Portoferraio with a medical friend called Emery. When he saw Napoleon next day he described a difficult and lightly guarded route over the Maritime Alps into the heart of France, and though it is unlikely that Napoleon then gave him any specific or possibly compromising instructions Dumoulin was so warmly received that he went back to organize a Bonapartist cabal in his home town and, with Dr Emery, to play a small part in opening the road to Paris in March 1815.

Another arrival from France was an old soldier who saw Napoleon in November and gave him an encouraging summary of French opinion. He said that he had been a month on the road, travelling by way of Lyons, Grenoble, Avignon, Toulon, and Genoa, and stopping at cafés and billiard-rooms along the way. 'They are waiting for you,' he told Napoleon; 'the present state of affairs can't last another six months.' Everybody was complaining about something and nobody was satisfied with anything. Even the people of Provence, who had been opposed to Napoleon and imagined 'that the larks would tumble into their mouths ready roasted' with the return of the Bourbons, had now changed their mind. Napoleon replied that he was ageing and needed a rest. In any case, he was said to have added, 'I have only one battalion and a corvette'. 'Not a soldier will fire on you', the *grognard* declared, 'and if an assassin were concealed in the ranks he would be torn to pieces.' But what of the risk of civil war? Napoleon asked, reminding his visitor that he had preferred abdication to fratricide. Nobody would fight, the man insisted; only the nobility and the priests stood to lose by the Emperor's return, and they were 'a handful of cowards who would not dare to show themselves'.

Other messengers seem to have been more ambivalent or less trustworthy. There was, for instance, an unidentified messenger who reached Portoferraio on 6 December. While the Oil Merchant observed that 'a good many officers said that members of the imperial family were very happy with the news' the man had brought, Peyrusse had a much less cheerful impression. 'From that day', he noted in his journal, 'His Majesty's character altered. His words were few, and he seemed morose and abstracted.' At the same time Napoleon sent the *Inconstant* to Italy to buy grain, strengthened the forts, opened new fields of fire for his guns, drilled his gunners energetically, and kept patrols on the move day and night. This informant, Peyrusse thought, was 'a foreign officer who had been in his service and had come from Vienna' to warn him of threats to his liberty or safety.

All through the months on Elba, moreover, it is clear that Napoleon was beset by Italian patriots urging him to put himself at the head of a popular rising, and he was justifiably afraid that they were indiscreet and that their movement was riddled with Austrian police informers. Whenever one of them came to see him he listened carefully, for all political intelligence was valuable, but he was equally careful to avoid being drawn into any futile plot. The case of Count Litta shows the need for such caution. Litta was a well-to-do Milanese nationalist, described by Lord Burghersh as being 'at the head of the discontented in Italy', and he arrived in Portoferraio on 30 November hoping to win Napoleon's support for a conspiracy that was now well advanced and also to engineer some rapprochement between the Emperor and Murat. Napoleon apparently asked how many foreign troops there were in the country and how they were disposed, whether the Italian prisoners of war had yet returned home, and how people generally were reacting to the Austrian occupation. But he was wise not to go any further, for Litta had crossed in company with the Oil Merchant, whom he took to be a fellow patriot, and on 30 December he told Mariotti's agent exactly what Napoleon had said. Within days Mariotti had given Campbell a full account of Litta's remarks, and two weeks later Lord Burghersh sent an official despatch from Florence to London which was clearly based on the same report.

Napoleon obviously had to run some risks of this kind, especially when amateur conspirators and enthusiasts turned up on Elba, yet the results were worth the risk. From his spies, relatives, correspondents, well-wishers, and visitors, as well as from the newspapers that came from all the main cities in Europe, he could get a good idea of the run of events. With such a flow of information, indeed, with such a vast knowledge of European politics, and with so much personal experience of the personalities with whom he was dealing, he was well placed to decide whether and when the moment had come for him to gamble again on Fortune's wheel.

13

CONFIDENTIAL FRIENDS

In April 1814, when the Tsar so irresponsibly installed Napoleon as King of Elba, he merely solved the most pressing problem created by the Allied victory, and within six months it was becoming clear to the Allies and to Napoleon himself that this act of charity had left his ultimate fate unsettled.

In a formal sense, it is true, the situation had not changed between spring and autumn. There had been nothing in the Treaty of Fontainebleau to suggest that his stay on Elba might be brief, or to permit any variation in the terms which had put him there. Whatever misgivings the Allied sovereigns had about his proximity to the mainland they believed that nothing could be done to alter his position unless he did something foolish, or unless he could be persuaded, as part of some general settlement coming out of the Congress in Vienna, to agree to move right away from the shores of Europe. In October, for instance, Castlereagh was suggesting that the Allies might buy an island in the Azores from the Portuguese and offer it to Napoleon as a permanent home, sweetening the proposal by generous compensation for the loss of Elba, and the Tsar dropped several hints that he could find a comfortable retirement in Russia. But it was an open secret that the French royalists had never been content with the Tsar's proposal. Talleyrand had objected while the treaty was still being drafted, and on 28 August the Duke of Wellington wrote to Lord Liverpool from Paris to say that Louis XVIII still 'wanted to send Bonaparte somewhere else than the island of Elba'.

That wish became stronger as the months passed. Louis wrote to Talleyrand on 21 October to say that he would be willing to go beyond the financial clauses of the Treaty of Fontainebleau 'if the excellent idea of the Azores were put into effect'; and in the course of the autumn Talleyrand took every chance that offered to suggest a new home for the Emperor, canvassing both Trinidad and St Lucia as suitable places to which he might be moved, with or without his consent, and noting the particular attractions of St Helena, lost in the remote wastes of the South Atlantic.

On 9 November, according to the Genevan delegate Jean Gabriel Eynard, the matter seemed settled. The King of Bavaria, Eynard noted in his journal, told him with delight that Napoleon was to be sent to St Helena, saying 'I would never have been easy as long as I knew that this devil of a man was so close to the Continent'; and ten days later the *Journal des Débats* in Paris also reported that a decision had been reached.

The rumours were not true, but they soon reached Napoleon. There was little that Talleyrand said in Vienna that did not come to the ears of one of his agents or sympathizers in the French delegation, and Lady Holland sent him a copy of the newspaper report as soon as she saw it; he was understandably troubled by this talk of finding him a new home, despite the treaty. He had hoped that the Allied sovereigns would leave him undisturbed on Elba, at least as long as the Congress lasted, or it suited his convenience, and he had so far assumed that he was in much less danger from his former enemies, who had up till now treated him with the respect due to a fallen emperor, than from fellow-countrymen who regarded him as a fallen tyrant.

★

There was certainly a real threat from France. Louis had already given proof of bad faith by failing to pay the pensions promised to Napoleon and his family, and several of his ministers made no secret of their hostility. But the situation was not quite as bad as it seemed, for Louis was trying to govern France by a curious system which was neither wholly autocratic, quite constitutional, nor really effective. His chief adviser, the vain and reactionary Comte de Blacas, was a disastrous choice, for Blacas was quite incapable of imposing any common course of action on subordinates who jealously ran their own departments without much reference to him or to each other; and though Louis himself was far from stupid, he let everything drift while the situation in France steadily deteriorated. Long before the end of 1814 his policy towards Napoleon had degenerated into little more than a malevolent muddle.

Some of the ministers were well aware of Bonapartist sentiment in the country and of Napoleon's restlessness on Elba. All through the summer Beugnot's police agents were sending him a flood of reports. A good many were silly, but others were more worrying. On 24 July, for instance, an informant in Nancy quoted what some of the townsfolk were saying about the former emperor. 'They speak of him as a man betrayed,' the agent wrote in a prescient comment. 'It seems that his reverses and mistakes have served only to soften public judgements on him. His follies, his outbursts of temper and the ridiculous side of his behaviour have only slightly undermined the blind confidence that the people and the soldiers have in his capacities. No one exactly conspires for him, but this stupid infatuation is itself a kind of conspiracy which must be a cause of concern since he can dream of profiting by it.' 'One can scarcely believe that Bonaparte has genuinely renounced any future for himself,' another agent wrote from Strasburg on 6 August; 'one fears that he is merely sleeping for a time, until France and Austria become involved in wars from which he can profit. . . . If the government becomes more reactionary, one day Bonaparte will return to France at the head of an army of malcontents who will overturn everything.'

Such remarks were common, especially from informants in the army, and they were matched by intelligence from Elba that should have been taken more seriously. On 9 September the Prefect of the Var passed on a report from an agent in Livorno. 'Everyone in Bonaparte's service, whether civil or military', the man wrote after a brief visit to Elba, 'is convinced that he won't be there long and that things will soon change in his favour.' Ten days later a colonel in the gendarmerie who had been to Livorno also came away with a similar impression. 'Bonaparte is very uneasy, very agitated, not daring to sleep in the same room for two nights running,' the colonel wrote; 'he is openly expressing his hopes of returning to the political stage. The accounts which portray him as showing indifference, resignation and a desire for tranquillity don't seem to be sincere, being rather a matter of politics on his part since, for the duration of the Congress, he must keep up a pretence which doesn't upset the powers on whom his fate depends.'

Beugnot knew the Bonapartes and his job. He had been a friend of Lucien, finance minister to Jerome in Westphalia, and Prefect of the Nord in 1813. Yet he seems to have been oddly ambivalent about these warnings of trouble to come. On 18 November the police in St Tropez arrested a women named Berluc who had come from Elba in a party of household staff laid off from the Villa Mulini for reasons of economy. She admitted that she was the mistress of a captain in the Guard, claimed that she was one of Pauline's ladies-in-waiting, and predicted that Napoleon and his men would soon leave Elba. 'I am convinced that a long period of repose is impossible for Bonaparte', Beugnot remarked when he reported this incident to Louis, 'and that once he has ceased to fear the Congress he will work out some plan to get away from the island and once more trouble the world.' At the same time he declined to take Madame Berluc's prediction seriously. 'She speaks of *his army*,' he wrote sarcastically. 'As if one could land in France with seven or eight hundred men, most of whom would desert as soon as they could!' It is not surprising that one of Beugnot's acquaintances was 'astonished how little importance he attached to what was happening in Elba, and how few means he had to keep watch on it', and although Beugnot sent two frigates to patrol the Tuscan sea when he moved on from the police department to become Minister of Marine he left four ships of the line, eight frigates, and twelve corvettes idle in Toulon.

Beugnot seems to have been afraid of being misled by sensational rumours or of creating unnecessary alarms. But some of his colleagues were irresponsibly casual. The Minister of the Interior, an intelligent but inconsequential man named Montesquiou, was so complacent that he did not bother to open some of the reports he received from the Prefect of the Var, who was increasingly worried that one spring day Napoleon might turn up in the south of France. The Comte d'André de Bellevue, who succeeded Beugnot as director-general of the police, was an ageing charmer who had quite lost touch with conditions in France during the twenty years he had spent in exile and, like Beugnot, he did nothing to get rid of officials who had served Napoleon and were in some cases still in touch with Fouché. He was so naive, in fact, that when he wanted a personal assistant he consulted Maret, who had been one of the Emperor's closest confidants, and then agreed to appoint a man called Monier, who had been Maret's own secretary. Once Monier moved into the police ministry its business was like an open book for Maret and the clique of Bonapartist intriguers who looked to him for leadership. And much the same was true of the Postmaster-General's office, where Comte Ferrand was surrounded by men of doubtful loyalty who so hampered his efforts to censor the mail that many Bonapartists felt that only minor precautions were necessary when they sent letters through the post. General Evain, for example, was an old friend of General Drouot who had stayed on at a senior post in the Ministry of War and he regularly sent messages to Elba by way of his sister, who was employed in the post office in Angers. The whole of the public service, indeed, was riddled with *attentistes*, waiting to see what would happen. Even if they did not give Napoleon active support at this nadir of his career, they had once owed their careers to him and they might well find themselves in his service again; and in the meantime they connived at the activities of his friends and inhibited the intrigues of his enemies.

Covert obstruction of this kind made it generally difficult for the Bourbons to govern France, and it presented particular difficulties in dealing with Napoleon and his supporters. Talleyrand, of course, would have been delighted to hear that his former master was deported or dead. His deputy, the Comte de Jaucourt, may not have connived at any particular scheme to kidnap or kill Napoleon but at the beginning of 1815 he began secret negotiations with Metternich about ways of disposing of both Murat and Napoleon with the approval of their royal masters. General Dupont, the Minister of War, had never forgiven Napoleon for disgracing him after the surrender of Baylen during the Spanish campaign, and he could count on the backing of the Comte d'Artois whose agents had been mixed up in the Maubreuil affair and tried to harass Napoleon on his way across France. The younger brother of Louis XVI and Louis XVIII was the Emperor's most bitter enemy, schooled in rough methods by the exigencies of the royalist resistance movement in which he had played a leading part, and of all the people close to the King he was the most anxious to see Napoleon removed from Elba, whatever the means employed.

Yet despite the malice of such well-placed men there was something curiously ineffectual about the series of French plots directed against Napoleon in the course of 1814. The orders were too vague, or they came too late. The agents were untrustworthy or incompetent, or simply unlucky. There was not enough money to do the job properly. All they did, in the end, was to alarm Napoleon and persuade him that he would eventually be transported or abducted from Elba, even if he did not first fall victim to a disaffected Corsican, a vengeful royalist, a turncoat with a grudge, or a French agent acting on instructions from Paris or Vienna.

★

Corsica was the obvious base for ill-intentioned adventurers. While many Corsicans would happily have given their lives for their island's famous son, many more would have been as happy to dispose of him. Most Corsicans had sided with the great patriot Paoli and the British against the French Revolution, or had become fervent monarchists, and both factions hated the Bonapartes with the vehemence of a vendetta. One of the most influential of these enemies was Pozzo di Borgo, the family connection and boyhood acquaintance in Ajaccio who had eventually made his career as an adviser to the Tsar and done his best to offset Alexander's inclination to deal leniently with the Emperor. Among the French troops who returned to Corsica when the British occupation ended in 1815 moreover, were a number of royalist officers who would have liked nothing better than to settle accounts with the Corsican Ogre. The Comte Chauvigny de Blot, for instance, was the assistant to the general who temporarily took command of the island, and he was in close touch with the Comte d'Artois. 'For the peace of the world', he remarked of Napoleon in the summer of 1814, 'this monster should cease to exist', and he soon made an effort to turn these aggressive words into deeds. Knowing that there were a dozen Corsican officers in the 35th Infantry Regiment which had garrisoned Elba under General Dalesme, and that there were other Corsicans living in and around Portoferraio, Chauvigny tried to infiltrate agents among them before Dalesme took his troops back to France. He also tried to insinuate spies and possible assassins among the Corsicans slipping off to join Napoleon on Elba. When nothing came of these amateurish schemes Chauvigny was recalled, but promoted and given another appointment.

Late in July, however, General Dupont despatched Louis Guerin de Bruslart to serve as military governor of Corsica. The timing was significant, for Talleyrand had just sent the Corsican royalist Mariotti to serve as his consul and confidential agent in Livorno, and the two ministers may have devised a scheme for concerted action against Napoleon from Corsica and Tuscany. Bruslart certainly had both the experience and the motive for clandestine operations against Elba. He had been fighting ever since the first battles of the revolutionary war. He had been active in the royalist *maquis*—he was a man, Napoleon said apprehensively to Campbell, 'who was employed for many years while in England in plots and conspiracies' and was capable of 'every act of personal hostility and oppression, even to the taking of his life'. And he had a personal reason to seek revenge. Ten years earlier, during the Consulate, Bruslart had persuaded another royalist officer to trust Napoleon's offer of a safe-conduct but on his return to France the man was seized and executed, and Bruslart had then sworn to kill Napoleon for this breach of faith.

Once Bruslart arrived in Corsica he should have been in a good position to carry out his longstanding threat. He had a large staff, which included men who shared his knowledge of irregular warfare, three regiments of troops, a powerful force of gendarmes, two frigates, a corvette, and a brig. While this force was not sufficient to mount an attack on Portoferraio, it was thought strong enough to defend Corsica if Napoleon should ever make a sudden dash across to his homeland, to maintain naval patrols between the neighbouring islands, and to cover any covert moves Bruslart might make against the King of Elba. But it was not as strong as it appeared. While the senior officers in the army had hurried to take their oaths to the Bourbons, most of the men and a good many of the junior officers were still loyal to Napoleon. Bruslart could neither persuade, bribe, or blackmail anyone close to the Emperor into a conspiracy, nor could he count on anyone except the ardent royalists among his own men for any desperate venture. He even found it difficult to maintain reasonable security about his own intrigues, for there was always a steady trickle of deserters and Bonapartist agents crossing to Elba and the crews of the warships at his disposal also appear to have been unreliable. In January 1815, when Napoleon was talking to Campbell about the two frigates which had lately begun to sail a patrol line close to Elba, he made the surprising claim that 'the crews of these ships were attached to himself, and gave him intelligence of everything which passed in the squadron'.

All the same Bruslart kept trying, and rumours of his plots were so common on Elba that it became almost impossible to distinguish fancy from fact. There was undoubtedly a serious alarm in the first weeks of September. The police suddenly became more vigilant, and Mariotti reported that it was difficult for anyone but English visitors to get permission to stay on Elba for more than two or three days. On 9 September, moreover, the *Inconstant* was fully manned at short notice and sent to sea with an additional complement of men from the Guard. Napoleon shut himself up for two weeks in the fortress at Longone, where he kept a sparsely furnished apartment, and visitors were kept away from him. After that time, Campbell noticed, he did not go out to take exercise 'except with four armed soldiers to accompany him'. But there was nothing to show whether it was caused by one of Bruslart's schemes, by one of Mariotti's men, or by the current threat from the Barbary corsairs. Napoleon himself was vague, telling Campbell on 16 September that 'a person of another nation had come to Elba as an assassin' and saying apologetically that this was the reason for 'some difficulties' which a few English travellers had experienced at the hands of the fiery-tempered Cambronne and the Portoferraio police.

★

Gossip about such plots was common enough in the hothouse atmosphere of the Congress of Vienna, yet one clandestine operation against Napoleon certainly came to the notice of the Austrian secret police in the third week of October. At that time they began to intercept a series of letters written in sympathetic ink which Mariotti had sent from Livorno to Talleyrand, to Comte Dalberg, who was a member of the French delegation in Vienna, and to General Dupont in Paris. In the first of these letters, dated 28 September, Mariotti regretfully explains that there is 'not much chance of carrying off Napoleon' because of 'the extraordinary precautions which he has taken against all strangers, especially those who arrive from France and Livorno, the continual changes in his domicile, and the hope among his soldiers of a change for the better when the Congress is finished'. He had therefore devised an ingenious scheme to seize the Emperor at a moment when he was separated from his faithful guards. Noting that Napoleon had already made two crossings to Pianosa on the *Inconstant*, Mariotti also reported that Lieutenant Taillade, who commanded the brig, was overworked, underpaid, and discontented with his lot. If the right man could be found to act as intermediary, Mariotti suggested, Taillade might be persuaded to take Napoleon prisoner on one of these visits to Pianosa and sail away with him to a prison island near Toulon.

The next letter, written on 3 October, was a cryptic progress report to Dalberg. 'I haven't forgotten the commission for Spanish wines,' wrote Mariotti. 'As soon as some suitably aged wines arrive I will procure some casks and I will send them to you in double tuns by way of Marseilles.' Two days later, in a letter which reported that he had despatched two more agents to Elba and was busily planning other forms of surveillance, Mariotti made the complaint so common among secret agents. 'I have done everything possible for success', he wrote dolefully, 'but I have been at great expense, and haven't yet received a single farthing.'

The Pianosa 'project to abduct the neighbour', as Mariotti described it, had been devised too late. On 20 September, over a week before Mariotti wrote to Vienna to say what he had in mind,

Napoleon had left his protected retreat at Longone, boarded the *Inconstant*, and gone off to Pianosa in an effort to settle the trouble with Major Gottmann. It is possible that he then decided that he had no more occasion to go to the island, if there was no future in his plans to colonize it. He may also have learned that he might be carried off during one of these visits. Whatever the reason, he never went again.

★

One way and another September was an anxious month for Napoleon. At the moment when Bruslart and Mariotti were actively plotting against him, and a host of lesser agents were trying to make trouble for him and worm out his secrets, he was unexpectedly faced by a threat from the Barbary corsairs.

These pirates, who sailed out of Tripoli, Tunis, Algiers, and smaller ports along the north coast of Africa, had long been a menace, and they had become bolder during the recent war, when the French and British navies were otherwise engaged, roaming all over the central Mediterranean to seize merchant ships and carry off their crews into slavery. Eventually, it was clear, the maritime powers would be forced to sweep the seas clean of them. In the late summer of 1814, however, the pirates were still marauding up the coast of Italy, and they seemed to be a direct threat to Napoleon. A corsair raid might be a clever means of capturing or killing him, or an Allied naval operation nominally designed to suppress the pirates could provide useful cover for a landing on Elba.

The first of these anxieties probably explains why Napoleon sent the *Inconstant* to sea on 9 September. Since two pirate vessels had been sighted close in to Elba he no doubt thought it wise to make a show of strength, and he may well have had advance notice of the news that Campbell sent to Bertrand from Florence next day. For Campbell had just heard from Admiral Hallowell that the Bey of Tunis had formally declared war on Naples and Genoa, and that 'cruisers from Algiers and Tripoli are at sea with similar orders'. This news was bad enough, for Elba depended greatly on ships trading from these two Italian ports. Yet the last sentence of Hallowell's despatch, written from Palermo on 31 August, was even more disturbing. The Bey of Algiers, he said, had gone even further, telling his ships to seize all vessels sailing under the Elban flag 'and the person of the Sovereign of that island also, should any opportunity happily offer of getting hold of him'. Campbell's reac-

tion to this ominous phrase was extraordinary. He simply suppressed it, and when he entered Hallowell's letter in his journal he
underlined the final words and gave a peculiar and scarcely convincing excuse for failing to pass on the warning. 'The last part,
marked in italics,' he wrote, 'was not sent by me to General Bertrand, as being personally offensive against Napoleon.'

Politeness could not have been the reason for concealing such
vital information from Napoleon, especially when Campbell was well
aware that he feared such a fate more than any other. It is more
likely that Hallowell's letter came as an awkwardly timed embarrassment to Mariotti's scheme for abducting the Emperor with Taillade's help. If Campbell had discussed the kidnapping plot with
Mariotti—and he had passed through Livorno just as it was being
planned—he would have been understandably reluctant to pass on
any warning which might deter Napoleon from boarding the *Inconstant* or sailing round to Pianosa later in the month. He still said
nothing about it four days later, when he saw Bertrand in Portoferraio; he simply presented a letter which contained a formal
notification of the Algerine declaration of war, and observed that
Bertrand 'seemed much agitated'.on being told that the British government had given no undertaking to protect Napoleon in such situ-
ations. Bertrand's alarm was understandable, especially if Napoleon
already knew from other sources of the personal threat which
Campbell had deliberately concealed, and this episode may have induced Napoleon to seek at least indirect help from his brother-in-
law. Campbell certainly had that impression. 'Murat's squadron is
frequently to the southwards of this island,' he noted on 22 October, 'cruising between it and the Bay of Naples, in order to protect
their trade against the Barbary Powers, but I do not learn that they
hold any communication with this place.'

Campbell clearly found it hard to make up his mind about the
corsairs. In September he assumed that they were hostile to Napoleon. On 24 October, however, when a Tunisian xebec anchored off
Longone, he thought its appearance might be 'a trick of Napoleon
to disturb this country'. There was a comic incident when the
corsair went ashore and sought an audience with 'the Great Lord of
the Earth', and though this request was denied as a precaution
against treachery Napoleon stopped long enough on one of his
walks for the captain to prostrate himself in greeting. This gesture,
the hoisting of the Elban flag, a salute of guns, an exchange of
gifts, and the prolonged stay of the pirate ship, which remained in

Longone for almost two weeks, all made the suspicious Campbell wonder whether there was some secret link between Elba and Tunis, or Elba and Toulon. 'It appears certain', he concluded in a long and confused memorandum on 12 November, 'that Napoleon has established himself on an amicable footing with this Power, or that he had bribed the captain of the ship with the advantage of sheltering in his ports, so as to be able perhaps to communicate with France.' Finding himself unable to answer his own conundrums Campbell then went off again to Livorno and Florence, claiming that he needed to consult the Tuscan police, and that he also needed to compare notes with Lord Burghersh, through whom he now sent his despatches to Vienna and London. He seems to have been disproportionately excited by the arrival of the Barbary corsair at Longone. He could not have made more fuss about it if he had suspected that Napoleon was planning an early escape in this very ship, or if—as he had also suggested—the corsair had come from Toulon with news of a Bonapartist conspiracy to seize 'that important place and the fleet'.

★

Campbell was anxious, for he had an uneasy feeling that if anything went wrong he would be blamed. Yet the journey to Florence marked a distinct change in his attitude to his mission on Elba. He had hitherto been conscientious. He now began to evade and even to neglect his duty, and the difference shows in his journal entries and in his despatches. Apart from the set-piece reports on his conversations with Napoleon on 31 October, 4 and 21 December, his reports consist of little more than the gossip picked up by his man Ricci and reports from the Oil Merchant which Mariotti passed on to him. By January, indeed, he was scarcely bothering to keep up with events, for he covered the whole month in a single diary entry of less than a thousand words and these few pages, like the surprisingly brief messages to Castlereagh, had to cover a visit to Genoa, the unmasking of the Italian conspiracy by the Austrian secret police, and a significant talk with Napoleon in which the Emperor spoke of his fears of deportation and his determination to resist such a move by force.

There is no doubt that Campbell had come to find life on Elba very tedious. 'Napoleon', he wrote peevishly in a note at the end of December, 'has gradually estranged himself from me, and various means are taken to show me that my presence is disagreeable . . . for the purpose of inducing me to quit Elba entirely.' He found these calculated snubs distasteful, and he almost let Napoleon tease him into resignation before he decided that he should retain his appointment until the end of the Congress in Vienna.

'I have resolved to make the sacrifice of my own feelings', Campbell wrote, adding the disingenuous comment that for reasons of health and amusement he would 'occasionally' be visiting Livorno, the baths at Lucca, and Florence. This was the opposite of the truth. Although Campbell had sought and had been refused permission to settle in Tuscany for the winter months, he none the less behaved as if Lord Burghersh had given him leave to confine his duties on Elba to short routine visits. He seldom went to Portoferraio, and even when he did go over for a few days he left as soon as he had seen and talked to Napoleon. This new arangement suited Napoleon very well. The more Campbell was away, the more the Emperor was free to see whom he pleased and to do what he wanted. It also suited Campbell, who much preferred the fashionably congenial company he found in Florence, especially the attractive Contessa Miniacci.

'No one knew much about her origins, or even what nationality she was,' the French agent Hyde de Neuville wrote after meeting her in November 1814: 'speaking all languages, knowing everyone in the world of politics, no one could tell what cause she served behind her mask of indifference. Campbell was one of her constant companions, and each time he came from Elba he brought her news of what was happening there.' Hyde implied that she may have been one of Napoleon's informants in Italy—and most useful, if that were the case, because she could report what Campbell was thinking and possibly influence his movements between Elba and Tuscany. But she may have been working for the Austrian police, or for the British, or have been a double agent or simply a woman who liked to give herself an air of mystery. Whatever role she was playing it seems that her charms were one of the reasons why Campbell now spent so little time at his post.

*

Hyde de Neuville, Campbell wrote in November, after Mariotti had introduced him to this notable French agent, 'was sent to Italy by the King of France's confidential friends, to collect information respecting Napoleon's situation and conduct at Elba', and Campbell seized this chance to tell what he knew about the island and to repeat his persisting worries about Napoleon's schemes and intentions. For all Campbell's strong personal desire to leave his post on Elba, his professional instinct told him that some day soon, in some fashion he could not predict, Napoleon would again make trouble for his old enemies, and he ran over the reasons for uneasiness that he had expressed time and again in his reports to Castlereagh. The King of Elba, Campbell said, had 'unlimited freedom of person and communication with the Continent ... was not sufficiently watched ... was still of a most restless disposition' ... and was quite capable of getting clean away from Elba if he chose to do so. He was delighted at the effect of these doleful comments, noting that Hyde 'took memoranda in writing ... and departed the following day in post haste to Paris'. Campbell and Mariotti, who had both become somewhat demoralized by their isolation and the apparent indifference of their masters to the threat from Elba, had for once found a sympathetic and influential listener. Campbell spoke 'with the greatest frankness', Hyde wrote afterwards, and could not have been 'more zealous', while Mariotti was 'an active and intelligent man of spirit, who inspired confidence'.

The possibility of an escape had been worrying Hyde ever since the Allies had put the Emperor in such an unsafe place. 'Don't you think that before Napoleon left Fontainebleau he looked at a map?', Hyde's old acquaintance Admiral Sir Sidney Smith asked him when he stopped in London in July on his way home from exile in the United States. 'Is the distance between Elba and the southern coast of France of any significance to a man who has marched from one end of Europe to the other? Wouldn't a few hours be sufficient for him to rejoin his troops? Smith's questions nagged at Hyde all summer, for he could not persuade himself that Napoleon's old supporters had really abandoned 'the man who had such magic power over them'. As long as Napoleon was so close to France and Italy, he wrote, 'we must realize that intrigue, ambition, party spirit and enthusiasm will all work for him, even without him ... conspiracy or no conspiracy, whether or not he is distressed by his situation, he is and will remain an immense danger'. On 20 August Hyde was back in London. 'My mission was nominally concerned with the suppression of the Barbary pirates in the Mediterranean', he said later, 'but the secret reason was to prevent Bonaparte attempting an escape.' The idea of combining an attack on the corsair ports with a descent on Elba certainly had the support of Smith, who had once been very active in clandestine operations along the Channel coast and had made a remarkable escape himself when Napoleon had shut him up in the Temple prison in Paris, yet nothing seems to have come of this scheme for disposing of the troublesome Bonaparte. 'The English will do little or nothing to keep him on his island,' Hyde wrote disconsolately to Louis XVIII when he got back from London. 'If he leaves, he can again overturn Italy or start a civil war in France. . . . I continue to believe that one must put the seas between him and Europe.'

When Hyde set off to Italy in October he was still using the marauding Barbary pirates as an excuse for his mission, but he was actually worried by the prospect of an Italian revolution, with Napoleon and Murat acting in concert and profiting from the mistakes of their enemies. If France were to give Bonaparte the money promised by the Treaty of Fontainebleau, if the Allies were to let Marie-Louise have the Parma duchies and allow her to return from Austria, if the foreign troops were to be withdrawn from Genoa, and if Murat were to be left undisturbed but unappeased on the throne of Naples, 'in eighteen months Italy will be ablaze'. Hyde saw the King of Piedmont, the Duke of Modena, and the Grand Duke of Tuscany, and in each place he heard the same story. The whole country, he reported to Paris, was 'a volcano', and Napoleon would soon return: 'his spies and supporters were everywhere'. In Hyde's opinion there was only one way to avert this disaster. Napoleon must be separated from Murat. He believed that a combination of pressures and inducements might persuade Bonaparte to leave Elba—a fresh start in the Americas might well seem a more attractive prospect than 'the hazardous adventure of a new conflict in France or Italy', and he was convinced that Murat was so anxious to stay on the throne of Naples that he would agree to any terms that left him there.

It was an ingenious idea. Metternich had already begun to think along much the same lines, for this kind of agreement would be a convenient way of stabilizing the situation in the Mediterranean if Austria and France were to find themselves allies in a war against Russia and Prussia. The duc d'Orléans (or some of his friends) may also have been interested. He was the only member of the royal family who was acceptable both to the old Jacobins and Bonapartists. He would have been the residual legatee if the Bourbon regime collapsed and Napoleon was too far away to seize his chance; and, no friend of Ferdinand, he could have accepted Murat and Caroline as the rulers of Naples. But before Hyde could make any headway with such a scheme he needed some sign of interest from Napoleon, and though he made several attempts to contact the Emperor there was no response from Portoferraio.

Early in November, after Hyde's disconcerting discussion with Campbell, and primed with fresh evidence with which he hoped 'to destroy the false security' of the French government, he went back to France, where he urged both Blacas and Beugnot to set up a stricter watch on Elba. Send two frigates and a despatch-boat from Toulon, he said in a letter posted urgently ahead on the road to Paris; change the garrison at Toulon, make sure that the troops in the southern departments of France belong to reliable units, replace doubtfully loyal officials in Corsica, make sure that French representatives in Italy are trustworthy, censor all letters between Elba and France.

It was all sound advice, and something was done. The two frigates soon came from Toulon, Campbell wrote afterwards, 'but the evil could not be averted by them, and they were of no use whatever'. As Bruslart, Mariotti, and Hyde all realized, the French royalists had been both hostile and irresolute towards Napoleon, and now time was beginning to run out.

14

THE THREAT OF DEFAULT

The easiest way for the French royalists to harass Napoleon was to default on the pensions which had been promised to him and his family when he agreed to go to Elba. After twenty years of exile the Bourbons were reluctant to subsidize the Bonapartes, and Louis XVIII could justifiably claim that he was himself hard pressed for cash. When he returned to Paris he found that the treasury was almost empty, that Napoleon had saddled him with a massive public debt, that he had to repay the loans made by the Allies during the long emigration, and that his supporters were expecting some tangible rewards for their loyalty. Yet his failure to pay a single franc of the money was a foolish act of spite. It should have been obvious to Louis that if Napoleon lacked the funds to maintain himself on Elba he would some day be forced to choose between bankruptcy and a desperate bid to restore his fortunes—and there could be little doubt what choice a man of his temperament would make. The Allies certainly saw the risk Louis was running. They not only thought his refusal to meet his obligations was dishonourable; they also believed that it was tactically most unwise to give Napoleon any excuse for claiming that France had broken the Treaty of Fontainebleau. Castlereagh and his colleagues in Vienna had already made several sharp enquiries about the pension before Talleyrand wrote to Louis on 13 October to say that the Tsar was asking 'disagreeable' questions about 'the silence of the budget on this matter'.

Strict observance was a point of principle with Castlereagh, and he was worried by one despatch after another in which Sir Neil Campbell (lately knighted) described Napoleon's growing frustration as the weeks passed without any mention of the pension. 'I have been informed from good authority that his present funds are nearly exhausted', Campbell reported on 22 October; and by the middle of November he was linking Napoleon's need for money with a warning about his restless intentions. 'If pecuniary difficulties press upon him much longer, so as to prevent his vanity being satisfied by the ridiculous establishment of a court which he has hitherto supported in Elba, and if his doubts are not removed', Campbell wrote, 'I think he is capable of crossing over to Piombino with his troops, or of any other eccentricity. But if his residence in Elba and his income are secured to him, I think he will pass the rest of his life in tranquillity.' In the next despatch, on 6 December, though Campbell was still saying that prompt payment might appease Napoleon, he ominously suggested that even prompt payment might not be enough if 'some great opening should present itself in Italy or France'. Only two days earlier, he added, Napoleon had been talking to him about 'the ferment which there is in France', and suggesting that if the discontent broke out into a new revolution 'the Sovereigns of Europe will find it necessary for their own repose to call upon him' to pacify the country.

★

There were several reasons why it suited Napoleon to plead poverty. The first was temperamental. His personality uneasily combined the natural parsimony of a poor Corsican soldier of fortune with the acquired largesse of an emperor who had ruled the greatest empire since the Caesars, and he had been as cheese-paring as an auditor at the same time that he was settling ducal estates on his marshals and drawing up plans to make Paris into the finest city in the world. He was just the same in exile. Both his mother and his sister Pauline gave him handsome subsidies while they were living in Portoferraio, yet he would pore over their domestic accounts to see what they were spending and whenever he ferreted out an item he thought extravagant—even the paint in Letizia's parlour or the blinds in Pauline's bedroom—he would scold the offender and re-coup the cost. Anything that made him fear that he might run out of funds intensified his quirky meanness.

The second reason was closer to the truth of the situation, though it was still not as pressing as Napoleon claimed. Even if he discounted once-and-for-all costs incurred in the first year, when he was rebuilding the Villa Mulini and San Martino, making a number of other improvements, laying in stocks and putting his little army on a regular footing, he could see that in the long run he could not make ends meet on the income from taxes, the salt pans, the tunny fishery, and the iron mines, especially if he proposed to keep up a military establishment which cost well over a million francs a year or more than twice his civil and household expenses. On a year-by-year basis he would have to supplement his revenue by something very close to the promised pension.

Thirdly, he hoped to make political capital out of his financial difficulties. The more he could demonstrate that his affairs really were in a precarious state because he had been denied his rights, the more he could excite pity for himself and contempt for the Bourbon government which had broken its word and sought to ruin him. This was not merely a question of public opinion, although he was too skilled a propagandist to neglect any opportunity of that kind. There was also a slim chance, in the topsy-turvy game the Allies were playing in Vienna, that it might suit one or other of them to use the dereliction of the Bourbons as an excuse for rehabilitating the Bonapartes. In a world mostly ruled by despots such chances did occur, and Napoleon had seen stranger turns of fate in the past fifteen years.

It was the fourth reason for Napoleon's anxiety about the unpaid pension that was the most important, and the most urgent, and it does much to explain why he was unwilling to sit patiently and watch his resources drain away. For him as much as for Louis, the money promised to the Bonapartes had become a symbol. If it was paid, he might consider himself both solvent and safe, for the payment would have been a token of recognition by the Bourbons and the Allies. If it was not paid, despite Allied protests, he would eventually face insolvency, deportation, or worse. It was a simple yet devastating conclusion. If Louis found that he could ignore the financial clauses of the Treaty of Fontainebleau, and that the criticisms made by the Tsar or Castlereagh in Vienna need have no practical effect in Paris, there was obviously very little apart from his own faltering conscience to stop the King of France flouting all the other stipulations, and treating Napoleon as an outlaw. That sinister possibility seemed even more likely after 18 December, when Louis approved a French government decree confiscating all the houses and other possessions of the Bonaparte family.

If Louis could seize Napoleon's property with impunity, why should he respect his person?

★

Napoleon began to make economies towards the end of September, when he was drawing up his budget for the coming year. By the standards on which the Allied sovereigns and their hangers-on were spending money in Vienna the savings he proposed were trivial, but in the cramped circumstances of Elba they were sufficiently striking to impress Campbell that he was becoming genuinely embarrassed for money.

To balance his accounts, Napoleon claimed, he would have to cut the number of servants at the Villa Mulini, pay half each official salary in promissory notes, reduce the mess allowances of his officers, reduce the supplies for his soldiers, dispose of obsolete cannon and other military stores, curtail the programme of road-building and other improvements, and even sell off a few public buildings, including a church, a barracks, and the town hall at Por-toferraio. 'So many people have been dismissed from his entourage', one of Beugnot's spies drily reported on 15 November, 'that he does almost everything for himself except go to market with his cook'; and John Cam Hobhouse, visiting Florence, was shown a letter from Campbell in which he said that Bertrand was so short of funds to run the royal household that he had recently replaced Napoleon's favourite Chambertin with the coarse local wine.

Such measures, of course, would also make sense if Napoleon had come to the conclusion that he would be leaving Elba before long, and that there was no need to waste money in the meantime. Another of Beugnot's agents, indeed, had remarked on this possibility as early as 8 October. Napoleon, he said, was 'reducing himself to the condition of a rich bourgeois, perhaps because he lacks the money to meet expenses, perhaps because he is uneasy about the result of the Congress and he wants to mask a change in his secret plans by affecting a misleading simplicity'.

It is true that there was a serious deficit on current account, and both the budget for 1815 and the final statement which Peyrusse drew up after Waterloo make it clear that Napoleon was living above his income all the time he was on Elba. But he was not yet living above his means. Few of those who heard his complaints of impending penury had any idea that the King of Elba still possessed substantial assets. It is difficult to be sure how large they were, because Napoleon was an adept at manipulating figures to suit his immediate purposes, and Peyrusse had long since learnt to balance the books to a few francs without revealing subventions, borrowings, and disbursements his master wished to conceal. In a round sum, however, it seems that Napoleon could still count on five or six million francs, possibly more. Of the four million gold francs Peyrusse had collected at Fontainebleau, Orléans, and Rambouillet before leaving for Elba, about one-third had been used to meet extraordinary expenses up to the end of 1814. That left over two and a half millions in the war chest. The reserves from the iron mines, which had been the cause of the quarrel with Pons, uncollected payments for iron ore, and deposits in a number of European banks amounted to something like another million francs (and bankers in Genoa and Geneva would have been willing to grant more credit on the security of the mines). Lavalette, who had been a close associate as well as Napoleon's Postmaster-General, said in his memoirs that ever since the Russian campaign he had been hoarding 1,600,000 francs in gold, packed in boxes made to look like fifty-four volumes of *Ancient and Modern History* and hidden under the floorboards of his house near Versailles; Prince Eugene and a few other trustworthy nominees apparently had money on call for Napoleon; and his mother had saved large sums of money from her official allowances during the Empire and, for all her Corsican frugality, was willing to put these at Napoleon's disposal for any great family purpose. Even if Napoleon had continued to run a deficit at the same rate as he had in 1814 he would have been able to maintain himself, his kingdom, and his army at least until the end of 1817. There was no immediate financial crisis. There was, however, a deteriorating political situation which had made the payment of Napoleon's pension the symbol of all his prospects.

★

Napoleon had been waiting for weeks for his sister to decide that she was well enough to return from the summer villa which Murat and Caroline had provided for her at Portici, in the shadow of Vesuvius. For most of October he had even forgone the use of the *Inconstant*, kept swinging at anchor in the Bay of Naples until Pauline stopped changing her mind about her date of departure. She thought she might travel by way of Rome, or take the waters on the island of Ischia, or wait to hear how much the Duke of Wellington had paid to turn her Paris mansion into the British embassy. At last Murat and Caroline saw her off on 29 October, and two days later the brig slipped in under the forts at Portoferraio and was greeted by a royal salute. Pauline had a strong sense of theatre as well as a good deal of egotism, and she knew how to make an entrance. Although Napoleon and her mother were on the quayside to greet her, with a barouche to take her round up the carriage-road to the Villa Mulini, she insisted on waiting until the sailors had unloaded her own *berline* and horses; in the meantime she made sure of an impression on the crowd by walking about and giving sweets to the rabble of children who pressed about her. That night the townspeople illuminated their windows in her honour.

Both Pons and Peyrusse noted the immediate difference that Pauline's arrival made to the King of Elba as well as to his subjects. For all Pauline's waywardness, her extravagance, her hypochondria, and her amorous instability, which made her more like a spoilt adolescent than a woman well past thirty, she had a disingenuous attractiveness which carried everyone along with her whims. 'The Princess Pauline had all the qualities of a consoling angel,' Pons said in his effusive way. 'She was sweet, affectionate and kind, and her gaiety gave life to everything about her.' Peyrusse said much the same. 'Her presence was a source of pleasure and enjoyment for the Emperor's court,' he wrote later, 'for the women of the town, and for the garrison.'

There was little anyone could do to compensate the Frenchmen on Elba for all that they had given up by following the Emperor into exile. Every day more soldiers grumbled that they were homesick, asking for leave or their discharges, and Campbell was only one of several who remarked on the poor morale of Napoleon's army and the way it was shrinking in size from one week to the next. 'We are bored to death', an officer told a visitor from France shortly before Pauline arrived. For such people Pauline's appearance was a symbol of hope. Like Napoleon's attempt to transplant the imperial court to Elba her effort to instil Parisian vitality into the drab small society of Portoferraio was largely make-believe, but it served a turn in the weary winter months when everything was unsettled and yet nothing definite could be done.

Pauline certainly enlivened the routines of the Villa Mulini, even at the expense of tiffs with her brother, who was irritated by her whimsies, her taste for flamboyant and immodest dresses, and her insistence on teasing anyone—including him—who was at all pompous. She had a stream of callers, whom she normally received as if she was a languid invalid, and when she went out to return the visits she had herself carried about in a sedan-chair. Yet she was always able to summon up energy for the entertainments she adored. There had been a ball to welcome her, where she danced half the night, and she soon followed this with a whirl of sociability. There were 'afternoons' or 'circles' for Elban ladies, musical evenings where the company was entertained by the Neapolitan sing-

Portoferraio from the anchorage. Fort Falcone is to the left, Fort Stella to the right, and the Villa Mulini stands on the ridge between them

Portoferraio: the harbour and the walls. The Water Gate is at the centre of the picture

Portoferraio: Fort Stella. The Villa Mulini can be seen at the left, below the fort

The house at San Martino, built for Napoleon at Pauline's expense

ers and a pianist she had brought with her from Naples, dinners at the Villa Mulini for local notables and visitors at which the talk was far jollier than Napoleon's standard combination of domestic trivia and military nostalgia, *soirées dansantes*, with party games in which Napoleon sometimes joined, to amuse the officers and ladies-in-waiting who made up the younger set Pauline attracted around her, and amateur theatricals on the tiny stage in the storehouse which Napoleon had converted into an assembly room.

Less than three months after Pauline arrived in Portoferraio, moreover, Napoleon carried out his plan to make a more substantial theatre out of a deconsecrated Carmelite chapel, lately used for army stores, financing the alterations by selling boxes to the more prosperous citizens and making the venture seem socially desirable by enrolling the subscribers into an academy with the motto *A noi la sorte!* ('We are the lucky ones!') Apart from Pauline's own theatrical amusements the Academy Theatre was to be used by travelling companies, and early in February 1815 Napoleon had even begun to estimate the costs of mounting a season of opera; if he could use military musicians, he calculated in a private memorandum, he might keep the costs down, and by the time he had done his sums to the nearest franc he concluded that he might actually make a small annual profit.

This frivolity was a welcome change for everyone in the court circle, and it also distracted attention from Napoleon's preparations for more serious business. It led Campbell to report to Castlereagh that 'the decrepitude of the Emperor increases rapidly', and by the time accounts of this sudden round of pleasure reached Paris and Vienna they had been exaggerated into suggestions that Napoleon was declining into early senility, playing practical jokes and running games with the guests at Paulin's parties. There was also more gossip about his amorous inclinations, for Pauline had taken on several attractive ladies-in-waiting and it was rumoured that she introduced more than one of them to her brother's bedroom as she had introduced women in the past. But Napoleon was essentially a prig about women, less seeking than sought after, and there is no evidence of any serious relationships during his months on Elba.

His personal affairs were a private matter, while the festivities were public and expensive. Even these modest festivities were an added cost at a time when Napoleon was complaining of poverty and cutting his budget for the coming year; and as the season of Carnival approached he felt obliged to put a curb on what could be spent. On 3 January 1815 he gave Bertrand instructions for a series of masked balls which would either be held in the assembly room at the Villa Mulini or at the Academy Theatre, and where Pauline's group of amateurs would perform two short plays called *Les Fausses Infidélités* and *Les Folies Amoreuses* as part of the entertainment. The numbers would have to be limited and there was to be strict control of the costs. 'The refreshments will be without ices on account of the difficulty of getting them,' Napoleon told Bertrand. 'There will be a supper which will be served at midnight. The total cost must not come to more than one thousand francs.'

The money grudged from Napoleon's privy purse was trivial. It would scarcely have paid for Pauline's costume at a New Year's Day party in the Tuileries a year or so before. Yet even the modest provincial style of these social occasions was proving a burden for the Elban guests, who had to buy ball gowns and fancy dresses, and meet a good many incidental expenses that Napoleon would not pay himself. Pons remarked that they were 'the road to ruin for the general finances of the town and for the particular finances of the more notable citizens', and he said that if Napoleon had not left Elba 'the moneylenders of Portoferraio would have finished by swallowing the patrimony of the propertied classes'. All the same, Napoleon's polyglot subjects vied for invitations—there were Elbans and Corsicans, Frenchmen, Italians, Poles and even a few visiting Englishmen on the lists which Bertrand prepared.

One Englishman, however, was made of sterner stuff. Captain John Adye, of the sloop *Partridge*, who ferried Campbell to and from the mainland and maintained a somewhat casual watch on Napoleon by putting into Portoferraio once a week when Campbell was away, declined his invitation for the gala opening of the new theatre on 22 January, for which Napoleon had brought over a company of actors from Livorno. 'I will explain that I am indisposed,' he wrote to his wife that morning. 'In fact I have no desire to go and be insulted by some French officer, as is their habit.' In any case Adye had more serious matters on his mind. 'Until it has been decided whether Murat will remain on the throne of Naples and whether Bonaparte will remain on the island of Elba', he had told his wife only two days earlier, 'Sir Neil Campbell and I will still have a worrying time.'

15

PERFIDIOUS INTRIGUES

If Napoleon seemed to be pressed for money even to put a festive gloss on his petty receptions, so was his father-in-law, on a far grander scale. For every franc that the King of Elba spent the Emperor of Austria flung away five hundred, and by the end of 1814 Francis had run up such debts to entertain the crowd of pleasure-seekers in Vienna that he lacked the ready money to pay his troops. If they had been forced to march against the Russians in the middle of the winter most of them would have gone without a coin in their pockets. On the day that Napoleon opened the Academy Theatre Francis organized a lavish Carnival party at Schönbrunn, with gilded sleighs, bands, cavalry outriders, and hundreds of torchbearers to light the guests home through the falling snow after they had been regaled with a sumptuous banquet and a performance of *Cinderella*. It was said that this event alone cost him more than half a million francs. And one such extravagance followed another, the brittle gaiety promoted by the Festivals Committee serving to conceal the tensions which had almost led to a new war and caused so much delay to a general settlement that the Congress itself had not yet been formally convened.

Marie-Louise watched the gorgeous procession of sleighs at Schönbrunn from her apartment window but—as the French minister in Vienna observed—'she was never present at any of the fêtes and daily gatherings'. She sometimes went in to the Hofburg, the imperial palace in Vienna, to watch a ball from a discreet place in the balcony, or attend a relatively private function, but for the most part she stayed out at Schönbrunn, where her routine was so uninteresting that the police inspector set to watch her eventually asked to be relieved of a task so tedious. She practised the guitar; and she translated *Ondine*. 'My stay here at Schönbrunn and my misfortunes have taught me to recognize my true friends', she wrote on 30 January; and Count Neipperg had now become the closest of them all. She waited for his daily appearance and kept him with her as long as possible.

Francis did all he could to alienate Marie-Louise from her

husband. She was referred to merely as an Austrian archduchess, and her son was brought up almost as though he were an illegitimate member of the royal family. That attitude, indeed, was accentuated by a Vatican ruling in November that her marriage to Napoleon may have been technically invalid. If that were the case she would have been no more than a royal concubine and her son a bastard— a regrettable situation that the Vatican proposed to rectify, now Josephine was dead, by a new religious marriage with the Emperor. But that was now the last thing that either Francis or Marie-Louise wanted, and the suggestion that the unfortunate alliance with Bonaparte might have been invalid all the time had obvious attractions for them both.

Some people said that these marital problems were the reason why General Köller was sent off on a mysterious visit to Italy in December. According to one of Baron Hager's agents he was going to get Napoleon's agreement to an annulment. But it is still not clear where Köller went or what he was doing in Italy. According to another rumour he went at Metternich's instigation to discuss Napoleon's voluntary removal to a safer place. A third story claimed that he had been despatched to conciliate Napoleon, perhaps to offer him a command if there should be a war with Russia. Napoleon was certainly intrigued to hear that he was in the vicinity. 'He seemed to me to view the report more with feelings of hope and eager curiosity than of apprehension', Campbell reported to Castlereagh on 21 December. Campbell himself was so sure that the Austrian general was on his way to Elba that he took the *Partridge* to Genoa in a futile attempt to meet him, and there was a persistent rumour on Elba that Köller had put in to Portoferraio and assured Napoleon that in certain circumstances he could count on Austrian and British support. Whether or not Köller's journey had anything to do with Marie-Louise's marriage, it was about this time that she removed Napoleon's imperial coat-of-arms from her carriage panels and made it clear to the French members of her entourage that (as Méneval put it) 'she had lost interest in her husband'.

With so much public interest in her affairs Marie-Louise was understandably uncomfortable among the kings and princes in Vienna, and eager to claim her new rights and title as Duchess of Parma. But even this request was long denied, with the connivance of her father. For no one in Vienna was keen to see her installed in such a strategic position, close to Elba, on the road to Venice and

Vienna, and with the King of Rome serving as a possible rallying point for Italian nationalists. There were specific objections from the French and the Spaniards, pressing the Bourbon case for the restoration of the duchies of Parma, Piacenza, and Guastalla to the Spanish princess whom Napoleon dispossessed to make way for his sister Elisa; and there was a growing reluctance to implement these and all the other clauses of the Treaty of Fontainebleau. The only person who spoke up for Marie-Louise, or seemed to have a good word to say for Napoleon, was the Tsar. 'The treaty is not being carried out', he said sharply to Talleyrand as late as 15 February, 'and we are bound to demand its execution; it is for us an affair of honour; we cannot depart from its stipulations in any way.' The Tsar's comment was primarily directed at the French failure to pay the promised pensions, but his longstanding dislike of the Bourbons made him go out of his way to support the claims of Marie-Louise.

His encouragement strengthened her resolution, and all through the winter she kept herself aloof at Schönbrunn. When she was offered Lucca instead she insisted that her motto was 'Parma or Nothing!' and submitted a long memorial to the Congress to make her case. 'My affairs are still not settled,' she wrote to the Duchess of Montebello. 'I am so troubled, sad, afflicted and misanthropic that my uncles compare me to a nun.' The comparison was scarcely appropriate. It was only with Neipperg's help that she got what she wanted. She was finally given Parma, on condition that she renounced the money and titles promised in the Treaty of Fontainebleau, and that her son remained in Vienna, a disinherited princeling and a hostage in whom both parents seemed to lose interest when the last hopes of a Bonapartist dynasty were dashed.

★

'We must hurry to overthrow Murat', Talleyrand wrote to Louis XVIII on 7 December, 'otherwise nothing will be done by February.' Talleyrand was increasingly worried about the stability of the Bourbon regime in France and he was anxious to get rid of all the Bonapartes as soon as possible. But he failed to make much impression in Vienna, since few of the Allied statesmen cared for the Bourbons. Apart from the British, who had given Louis asylum, they had all at one time or other been Napoleon's allies and they were well aware that in the event of another conflict he would prove to be the most valuable asset to any sovereign who could enlist him, either as the most brilliant field commander in Europe or at least as a political force to embarrass an enemy; and although Murat was obviously less formidable he too might be turned to similar account in a crisis. There were suspicions, indeed, that the British might already be playing a double game. One of Beugnot's agents suggested that they might be holding Bonaparte as a hostage to ensure that they would get what they wanted at Vienna, and though such notions were contrary to obvious British interests, the idea that Britain was capable of perfidious intrigue persisted, and was strengthened by Lord William Bentinck's encouragement of Italian nationalists.

There is no evidence that the British government ever contemplated an understanding with Napoleon, or the maintenance of Murat on his throne. The Tsar, however, had vacillated so much in his attitude to Napoleon that he could easily change his mind again if the crisis led to a new war; for Napoleon's presence on Elba would then be a distraction to the French and a threat to the Austrian rear in Italy. His open patronage of Prince Eugene was one way of hinting at that prospect, and Talleyrand passed on an even more explicit remark when he wrote to Louis on 12 November. 'Austria thinks itself sure of Italy', the Tsar was supposed to have said, 'but there is a Napoleon there who might be made useful.' Louis was both impressed and worried. 'I credit the report', he replied to Talleyrand ten days later.

Murat could also be mobilized against the Austrians—even helped against them if the need arose, for the Russians had a base as close as Corfu. This threat alone, given the rebellious mood of the Italians and Austria's inability to reinforce the army in Italy while there was a risk of war over Poland, was enough to make Metternich reluctant to quarrel with the King of Naples. Metternich was even afraid that Murat might cause trouble without Russian encouragement or without being provoked by the rumoured French plan to send a naval expedition to dethrone him and restore the Bourbon king in Naples. For Murat was restless and fearful. 'I have begged the Queen to calm him, to engage him to stay peaceful, not to ruin his affairs by some *faux pas*, and to await the decision on his future', the Austrian ambassador de Mier wrote to Metternich from Naples on 12 November. If the Allies should finally decide to throw him over, de Mier added with understandable anxiety, 'you must expect that he will have recourse to every imaginable means to maintain his position, and that his armed uprising would bring incalculable misfortunes'.

For the time being, in fact, it was mutually advantageous for Metternich and Murat to observe the treaty which they had negotiated in 1814, although neither party trusted the other and the treaty had still not been formally ratified. On the one hand it gave Metternich some guarantee that the hot-headed turncoat would not raise all Italy with the cry of national unity and freedom against the unpopular Austrians. On the other hand it gave Murat some assurance of Austrian protection as long as he did not give Metternich an excuse for repudiating the agreement. Metternich also counted on the astute Queen Caroline who (as Talleyrand reminded Louis on 25 September) had been Metternich's mistress when he was the Austrian envoy in Paris and had always kept in touch with him. Metternich was sure that she knew the value of the Austrian connection and would do all she could to prevent her husband risking his throne and his neck on some impulsive escapade.

A man of charm, with a nature that combined passion and intelligence, Metternich had often turned his relationships with women to advantage, but the unfortunate effect of his obsession with the beautiful Duchess of Sagan had been the talk of Vienna in the critical weeks before Christmas. 'He has not had his head about him for the past several weeks, so much has his thwarted love affair with the Duchess of Sagan absorbed him', Friedrich von Gentz was reported to have said to the King of Prussia; and when the Danish foreign minister noted this piece of gossip in his diary he added that Gentz believed 'it was for that reason that affairs had not moved ahead and had been so badly handled'.

★

Suddenly things changed for the better for Metternich, at least politically. The Prussian demand for Saxony had made it impossible for Castlereagh to combine the German states against Russia, and thus to force the Tsar to give ground in Poland. By the end of 1814 his policy had actually been turned back to front, with Prussia and Russia making common cause against Austria. But this threat had produced a new alignment, driving Castlereagh towards a close understanding with Metternich and Talleyrand. The danger of war was so great in the last days of the year that the three men held a series of covert meetings over Christmas, and on 3 January they made a secret alliance. In the few weeks since the statesmen first met in Vienna the situation had so altered that France had been totally rehabilitated. A defeated and occupied country had become an important ally whose troops might be called upon to restore a balance of power in central Europe. The alliance was thus a relief for Metternich, a diplomatic victory for Talleyrand, and a *pis aller* for Castlereagh.

At the same time, when war seemed likely from one day to the next, Castlereagh unexpectedly found himself in a much stronger bargaining position. On New Year's Day a courier rode in from Ghent to report that, after almost a year of haggling, Britain and the United States had at last signed a peace treaty and the long distraction of the American war was over. It was, Castlereagh wrote to the Prime Minister in London, 'a most conspicuous and seasonable event, which has produced the greatest possible sensation here', and within a week, and only two days after he had initialled the secret alliance, he could cheerfully add that 'the alarm of war is over'.

Napoleon and Murat had been relatively safe before the crisis, when everything was so uncertain, but the good news from Vienna was bad news for them. The Tsar, for instance, had now lost interest in them as potential sources of trouble for Austria; and at the same time Metternich no longer needed to keep Napoleon close at hand in case he was suddenly required to drive the Cossacks away from the gates of Vienna, or to placate the King of Naples in case he started a war by sheer impetuousness. On the contrary, as Russian pressure began to push him towards the secret alliance with France, Metternich realized that the ultimate removal of Napoleon and Murat was part of the price he would have to pay the Bourbons for their support against the Tsar.

But the problem of Naples had become so delicate and complicated that Metternich was not prepared to discuss it in detail with Talleyrand. He was far from trusting a man with such a devious past, and he was convinced that there were informers in the French delegation who were reporting to Napoleon. He had therefore opened direct negotiations with Paris, using Count Bombelles (a French exile in the Austrian service) and Blacas as the links between himself and Louis.

The first contacts had been made in November. Then, on 21 December, Bombelles saw Blacas and admitted to 'a lively Austrian interest' in the problem of Naples. Blacas, in return, declared that the situation in Italy was 'the only impediment' to closer relations between France and Austria, and that Louis would never feel safe on his throne 'so long as Murat offers asylum to every malcontent in the country, and is able within forty-eight hours to put his brother-in-law at their head'. The King, Blacas added with emphasis, 'would happily accept any sacrifice to get rid of Joachim of Naples'. What he wanted, in effect, was an Austrian agreement to disregard the treaty with Murat if the French and Spanish Bourbons joined forces to attack him. On Christmas Eve 1814 Bombelles sent this vital message to Vienna, using his valet as a courier. Metternich did not reply until 13 January, when he sent two despatches to Paris, one of them so confidential that Bombelles was told to read it to Blacas without leaving a copy, in which he assured Louis that the Emperor of Austria understood his desire to see a Bourbon back on the throne of Naples but insisted that the French must not be over-hasty. 'Once stability has been restored to Europe', he said, 'the fate of King Joachim will not be a problem.'

Metternich was not to be hurried, especially when the ending of the crisis had removed one of the main reasons for urgency. He was a natural mediator, who greatly preferred diplomacy to the harsh arbitraments of war, and his first impulse was to induce Napoleon and Murat to transport themselves peaceably and with adequate endowments to more distant domains. He also disliked the idea of French intervention in Italy, for he considered the whole peninsula to be part of Austria's sphere of influence and he was far from eager to see the Bourbons regain control of Naples. Above all, he was afraid that the French might act rashly and ineffectively. If any force was to be used, then it should be used by the Austrians, for Baron Hager's secret police had many reports that the French army was so sullen and disorganized that it would be very risky for Louis to put his troops on a war footing, let alone move them to Elba against Murat. Lord William Bentinck had come to the same conclusion. Given the present state of the French army, he told Castlereagh in late December, 'it would be dangerous to assemble it anywhere, for any purpose'. If such disaffected units were to be marched through Austrian-held territories in order to come at Murat, as Louis was now proposing instead of his original plan for a combined French and Spanish attack by sea, they could well spark off and join an Italian revolution. 'And what role will Napoleon play in all that?' Metternich asked anxiously, as he saw such a move providing the King of Elba with an unparalleled opportunity to cause trouble. If French divisions ever reached Piedmont or Lombardy he might well cross over to take charge of them, drive the Austrians out of Italy as he had done at the beginning of his career, and after Marengo, and arrive at Vienna to demand the restoration of his wife and child.

While these secret negotiations were proceeding Metternich was not willing to let Talleyrand jeopardize other vital Austrian interests by raising the embarrassing question of Murat's future: 'it is not a matter for the Congress', he said firmly, asking Blacas to instruct Talleyrand accordingly. The French must realize the constraints which the treaty with Murat imposed on Austria, and appreciate the dangers of an ill-prepared attack on him. If they would wait patiently to the end of the Congress, he concluded, 'the business will settle itself', and Austria would then be glad to help in clearing up the last vestiges of Bonaparte's empire.

The French were not satisfied with this veiled offer of Austrian collusion in Murat's overthrow, which left the moment and the means of it uncertain, and they were anxious to extract the best possible terms from Austria while the signatures were scarcely dry on the new treaty of alliance. Blacas therefore increased his pressure on Metternich. In two interviews with Field-Marshal Vincent, the Austrian ambassador in Paris, who knew nothing of the clandestine correspondence and confessed to Metternich that he was mystified by some of the things that were said to him, Blacas urged that Murat should be dethroned as soon as possible, and that order should be restored throughout Italy. 'If nothing is done', Blacas told Vincent, 'then one day we shall see the man from the island of Elba appear in Italy, menacing the security and disrupting the tranquillity of France and Europe.'

Italy! The misleading conviction persisted in almost every prediction of danger. 'It may be expected, should Murat disappear,' the British agent Fagan wrote to Lord William Bentinck about this time, 'that we may soon see the Emperor of Elba at the head of Murat's army'; and it suited Blacas to exploit such fears even if this meant neglecting the threat to France. He first made Vincent feel uneasy. He next tried to draw the Austrians into a definite plan to destroy Murat. At a meeting with Bombelles on 31 January he said that he was now willing, as Metternich had requested, to tell Talleyrand to stop talking about Murat in the Vienna meetings, and that he was also willing to moderate his demands for immediate action—but only on one condition. The Emperor of Austria must sign a secret convention recognizing the rights of Ferdinand IV to the throne of Naples, promise his help in recovering it for the Bourbons, and agree to complete the operation no later than six months after the end of the Congress.

That concession was essential. It would give Metternich time to settle other matters which he considered more pressing than the future of Murat. It would allow him to consult the Russians and the British, who were bound to have strong views about any changes in southern Italy, and about any 'adjustments' in Austria's favour in the rest of the country, and it would relieve him from the immediate French pressure for a punitive campaign which would mainly benefit the Bourbons and might well be very damaging for the Habsburgs. All the same, he would have to pay a stiff price for it, since Blacas was well aware how much bargaining power France had already gained from the Russian threat of war over Poland and Saxony. France would wait, he said, but only if the Emperor of Austria accepted the principle at once, and recognized it in a binding if secret agreement.

Louis was now so keen to act against Elba and Naples that he had lost sight of the possibility that his government might not survive to act at all.

16

NECESSARY PREPARATIONS

The stage was now set for the drama to come. As Napoleon cast his balance sheet of dangers and opportunities at the beginning of 1815 he could only draw one conclusion. One way or another, his fate would be settled before he had been King of Elba for a twelve-month.

In the first weeks of January, of course, he did not know that the risk of war was over, that the Allies had begun to deal with their outstanding business, and that Blacas and Metternich were now actively conspiring against him and his brother-in-law in Naples. But the pressure on him had been growing all through the autumn, as the boredom and the pettiness of his daily life became more un-bearable, as one plot succeeded another, as he saw that Louis had no intention of paying his pension, as he learnt of Talleyrand's hos-tile intrigues in Vienna, and as he came to realize that the Treaty of Fontainebleau—which none of the signatories except the quixotic Tsar appeared to take very seriously—was his only guarantee of protection apart from Campbell's ambiguous presence and the loyalty of his Guard. If he stayed on Elba, even for a few more months, he was finished. At best he might be offered a bribe to take himself off to some final place of exile, and at worst he might be transported, imprisoned in a fortress, or killed defending himself.

His mind was running on such grim possibilities when he talked to Campbell on 14 January about the recent strengthening of the defences on the land approaches to Portoferraio, and he seemed 'much agitated' by the appearance of the two French frigates which Beugnot had recently sent to cruise between Tuscany and Corsica. Beugnot intended the *Melpomène* and the *Fleur de Lys* to act as a deterrent to escape. Napoleon took a different view. As the French warships flaunted the Bourbon flag along his northern coast he claimed indignantly that they had been stationed there to cover an attempt at assassination or abduction. 'Not by the English,' he told John Macnamara in a long talk at San Martino on the previous day, 'they are not assassins. I am obliged to be cautious in regard to some others, especially the Corsicans, some of whom have strong feelings against me.'

The rest of the four-hour conversation with Macnamara, however, was very like all the other chats with visiting Englishmen. Napoleon talked about his defeat in Russia, his projected invasion of England, the disastrous effect of Marmont's desertion, and even such personal matters as his health, the reported suicide attempt at Fontainebleau, and his separation from his wife and son. 'What, kill myself?' Napoleon asked. 'Had I nothing better to do than this— like a miserable bankrupt, who, because he has lost his goods, determines to lose his life? Napoleon is always Napoleon, and always will know how to be content and bear any fortune. It must be confessed that I am in a better plight now than when I was a lieutenant of artillery.' The only hint that he might be thinking of yet another change of fortune came when he asked Macnamara about the state of France. 'We had a storm last night,' Macnamara replied; 'now there is no wind, but the sea is agitated.' Napoleon liked the simile. 'Well answered', he said.

The constraints and threats of the past few months had accentuated his natural restlessness, and his still unsatisfied ambition was undoubtedly making him look for a way of playing some new role in the affairs of Europe. He could see the prospects opened by the ferment in Italy, by the increasing unpopularity of the Bourbons among the classes which had benefited most from the Revolution and the Empire, and by the evident disaffection in the French army; and though the chance of profiting from the disagreements among the Allies was slipping away faster than he could tell at such a distance from Vienna, there was always the hope that they might let him return in peace, and of defeating them if they preferred to fight. If he left Elba, with his luck and the advantage of surprise, he would at least have a run for his money, and he might do far better than that.

Some people, indeed, thought that Napoleon had already made up his mind. 'Everyone believes that the revolution will break out in March', the Oil Merchant wrote from Portoferraio on 6 January 1815, 'and that all the necessary preparations have been made.' He had no real evidence to support this belief. He had been hearing much the same gossip ever since he first landed on Elba, and he knew that his friends in the garrison so ardently hoped that Napoleon would lead them back to Italy or France that the wish was possibly father to the thought. Yet when he returned after a three-week visit to the mainland he sensed a definite change of mood. It stemmed in part from the fidgets of the *grognards*, veterans who could catch the first sniff of a new campaign. The officers, too, seemed to think that something was in the wind at last, and people who went regularly to the Villa Mulini felt that the Emperor had the air of a man who knew what he proposed to do and was simply waiting for the right moment to do it.

Yet nothing was actually settled by the middle of January—certainly not a specific date, a means, or a destination. Napoleon was a natural improviser, and so much depended upon events beyond his knowledge and control that he could do no more than prepare to act quickly when he received a peremptory warning to flee or saw some dazzling opportunity to strike. He could accumulate stores and munitions, concentrate his troops round the harbour and exercise his gunners on the pretext that he might be attacked. 'The troops at Portoferraio have received the order to be ready to march at short notice', Bruslart reported to Paris on 8 January, and two days later he added 'I have been told that there is a very active state of security at Portoferraio, and that there are patrols day and night'. Napoleon could even get his ships into seaworthy condition, for no one could tell whether he was planning to repel a hostile squadron or to sail away in his own little flotilla. In his uncertain situation readiness was all.

★

Suddenly luck turned against him. In the early hours of 13 January the *Inconstant* was wrecked on the far side of the roadstead at Portoferraio. The brig was essential, whether he stayed on Elba or proposed to leave the island. For eight months she had been his main and most dependable link with the outside world. She was such a familiar sight, indeed, that the French and British men-of-war patrolling off Elba had come to take her erratic voyages for granted. *Inconstant* was not much of a warship, it is true, for her poorly trained crew and incompetent officers could scarcely handle her properly, and they could not have stood more than one broadside from a frigate. But she was sufficiently seaworthy to risk a voyage to America, if Napoleon decided to seek asylum in the United States, and she was large enough to carry a battalion of the Guard on the relatively short crossings to Italy and the south of France. At such a critical moment her familiarity would be a great asset, giving Napoleon some chance at least of getting away on the brig without being stopped and searched. He had managed that trick once before, when he had given Nelson the slip and sailed unchallenged all the way back from Egypt, and he might well expect to do the same again. It was, after all, less than two hundred sea miles from Portoferraio to that familiar beach at Fréjus.

The wreck of the *Inconstant* was a curious business. On 3 January the brig called at Civita Vecchia to collect Ramolino on his way back from a confidential journey to Rome and Naples. The weather was already bad and next day, when Lieutenant Taillade left for home, the easterly wind was so strong that he was driven clean past Elba and round the tip of Corsica into the safe naval anchorage at St Florent. It was an embarrassing landfall, for Taillade had no business risking Napoleon's only substantial ship in Corsican waters, especially with Ramolino on board and liable to arrest if he was discovered. Since there were already doubts about Taillade's loyalty he would have been wise to keep well clear of the island which was governed by the vengeful Bruslart.

Yet Taillade seemed oddly unconcerned, and he stayed on for several days, claiming that he was delayed by adverse winds and the need to repair his damaged rigging. During that time he talked quite openly with Bruslart's assistant, Colonel Perrin, with the captain of the frigate *Uranie*, which had been brought up close to keep a watch on the *Inconstant*, and even with Colonel Pivet de Boessulan, a fervent royalist who was suspected of having landed on Elba to make an attempt on Napoleon's life. It was compromising company, and it was said afterwards that some of the talks were in English (Perrin had been in England as an exile and Taillade as a prisoner-of-war) so that Ensign Sarri, Taillade's second-in-command, should not understand them, and that Taillade was offered a commission if he would desert to the Bourbons. If Taillade was tempted, and indiscreet, however, he was not a man for bold actions, and on 10 January he took advantage of a west wind to run for Elba where Napoleon was waiting anxiously for news of his missing ship.

Conditions were still bad. Captain Adye noted in the log of the *Partridge* that his sloop had lately been buffeted by 'strong gales, snow and sleet', and in Portoferraio the Oil Merchant vainly tried to get a passage to the mainland. 'Foul weather stops me from sailing', he noted on 10 January. 'Blown back into harbour', he wrote next day, but some vessel had evidently reached Elba for he added that 'a courier has arrived from France with news of unrest'. Late on the following afternoon the *Inconstant* was caught by strong winds from the north and Taillade ran down from Capraia in darkness and heavy seas. With more luck than judgement he somehow scraped through the narrow and dangerous passage between the Portoferraio lighthouse and the Scoglietto rock, tried to come round in the lee of Fort Stella, and was caught in irons. With the gale driving him backwards across the harbour he put down two anchors, but they soon dragged and towards dawn on 13 January he was drifting towards the small sandy beach between the rocky points at Bagnaia, firing cannon as a signal of distress. By the time Napoleon had been roused and had ridden round the bay the *Inconstant* was already stranded, with the waves breaking over her and her rigging in a terrible tangle. For some hours it seemed that the brig might be a total loss, but as the storm subsided it was possible to lighten her by shedding guns and cargo, and she was eventually pulled off the sand and towed across to the main harbour.

Was the *Inconstant* run ashore by design or accident? Campbell soon heard rumours that Taillade had done it deliberately. 'Some people say', he wrote a month afterwards, 'that Napoleon suspects him of a secret understanding with the existing Government of France, and of a wish to destroy the brig.' When Drouot conducted an inquiry into the disaster, however, Taillade was merely charged with 'incapacity and peculation'; and though he was obliged to turn over his command to a newcomer from Toulon named Chautard, he was nevertheless retained in the Emperor's service. Peyrusse later put forward the more charitable but bizarre theory that the *Inconstant* had actually been wrecked on Napoleon's orders, so that the brig could be secretly refitted and restocked while she was ostensibly under repair. But both these notions were too flattering to Taillade's skills as a sailor. Even if Bruslart's men had suborned him during his visit to Corsica, or Napoleon subsequently took advantage of the repair work to smuggle military supplies on board, Taillade obviously lost control of his ship as he came into the Portoferraio roadstead that night, and he and all his men were lucky to survive.

Whatever the cause of the accident its consequences could have been serious for Napoleon. If the *Inconstant* had been a total loss he would have been deprived of his one sure means of escape, and in any case the damage could have made the brig unserviceable just when he needed it most. Without her his sole chance of getting away in an emergency would be a desperate flight in a small boat, such as the *Caroline* or the *Mouche*, with only a corporal's guard to protect him. Yet Napoleon was lucky. Since it would take no more than two or three weeks to get the *Inconstant* ready for sea again, he could count on the vessel being available by the middle of February.

This unfortunate accident greatly helps to date the moment when Napoleon finally made up his mind to leave Elba. He could never have planned to go in January, when the weather at sea and in Europe was bad, when he had scarcely begun to prepare for such a venture, and when the situation in Vienna and Paris remained so confused that he could not have come to any conclusion about his future. In any case, if he had seriously considered an escape sometime in January he would never have sent the *Inconstant* away to Civita Vecchia: he would have wanted her close at hand. If he had thought of escaping in early February and proposed to take his troops with him, he would have had to set things in motion in the middle of January, for he needed about three weeks to mount his little expedition, and by that time he knew that the *Inconstant* was out of commission. But a decision reached in early February fits all the known facts, including Bertrand's later remark that there was no question of any plan for escape until three weeks before Napoleon sailed for France. Everything and everybody could be ready by the end of February. The *Inconstant* would be repaired, stores packed, troops alerted; and there would be a series of dark nights on which Napoleon might have a better chance of evading the *Partridge* and the patrolling French frigates.

The wreck had been a worrying nuisance. But by making Napoleon realize how much he depended on the vessel for survival, or escape, it may well have added to the growing list of reasons for him to get away from Elba while the going was good.

*

As January turned into February there was no apparent change in the pattern of life in Portoferraio. The series of balls went off well, culminating in the fancy-dress procession which marked the end of Carnival on Ash Wednesday. Napoleon himself seemed preoccupied, but Pauline gave some sparkle to the events, and there was still a steady flow of people crossing to Elba, as Campbell put it, 'from motives of curiosity and speculation'. Napoleon was making a successful attempt to keep up an appearance of normality, and it was lucky for him that the Oil Merchant and Campbell—the two agents on whom surveillance most depended—were both away for much of the time. The Oil Merchant, who left on the day that the *Inconstant* was wrecked, was gone for almost five weeks, and there is nothing to show why or where he went; while Campbell was now carrying out his plan to spend most of his time on the mainland and to make a brief visit to Elba every fortnight. He only saw Napoleon on 14 January and 2 February, and then on 14 February he had to be satisfied with a long and rather disagreeable interview with Bertrand. Between these short visits he had to rely upon Captain Adye to keep some sort of watch on Portoferraio harbour, and on his informant Ricci to pick up the gossip of the town. Such casual surveillance was almost worthless.

Campbell plainly suspected nothing when he saw Napoleon on 2 February, for he described the Emperor as 'unusually dull and reserved' and he could find nothing interesting to say about their meeting. There was a good reason for Napoleon's sombre mood. Three days later, when the repairs to the *Inconstant* were completed, he sent Pons a confidential message, asking for a report 'on the means of organizing an expeditionary flotilla' and on the following day, when he talked directly to Pons, he glossed that intriguing request with a revealing question. 'Shall I listen to the wishes of the army and the nation, who hate and mistrust the Bourbons?' Pons diplomatically replied that the Emperor's return to France would be a source of great happiness, provided that it did not lead to war, but he then guessed that Napoleon had made up his mind to escape and he was delighted that he was the first to know. Such a confidence, he boasted in his memoirs, was 'an honourable distinction', and Napoleon comfirmed that claim. 'Pons alone knew the truth,' he said afterwards; 'neither Bertrand nor Drouot was in the secret of my return; I confided only in Pons because his cooperation was essential to the preparation of the necessary ships.' Yet Peyrusse also took a hint. 'In the first days of February', he wrote, 'His Majesty asked me for 500,000 francs, and on the same day I was told to go and deposit my cash-box in Fort Stella. I knew enough to have an inkling of the reason for this removal. So in great secrecy I put by some flour, some wine, some potatoes and salt beef, and I waited to see what would happen.'

What prompted Napoleon to give these preliminary orders to Pons and Peyrusse at the beginning of February, when he had apparently done nothing to alert his military subordinates? In later years he claimed that Murat had misled him by sending Colonna d'Istria from Naples to say that the Powers had finished their work in Vienna and that the Tsar was on his way back to Russia. 'I left the island of Elba too soon,' Napoleon said to Colonel Gourgaud on St Helena. 'I believed the Congress to be dissolved.'

It was easy to blame Murat after the event, when he was dead and Napoleon had good reason to regret the untimely and even unwelcome assistance of his brother-in-law, but that particular excuse for the disastrously timed decision to escape does not fit the dates. Colonna did not leave Elba for Naples until 7 February, which was two days after Pons had been told to work out how the Guard could be shipped to France, and there can be no doubt that Napoleon had sent his mother's chamberlain to give Murat and Caroline some idea of what he had in mind. 'Colonna will say a great many things to you which are momentous and important', Napoleon wrote in a covering letter which shows that he already had a departure date in mind. 'I count on you, and above all on the greatest celerity—time is short.' It is of course possible that Colonna brought a false report about the Congress when he returned to Portoferraio on 11 February, but by then Napoleon was working to a timetable and any message from Murat would merely have confirmed his intention to leave.

If there was a specific report about the Congress which brought Napoleon to the point of decision in the first week of February, it is more likely that it came directly from sympathizers in Vienna than from Naples. There was, for instance, a persistent rumour at the time that two English aristocrats had made a special journey from Austria to warn him that he would soon be transported, and when the Oil Merchant eventually came back to spy in Portoferraio he passed on a similar story. 'About twelve days ago', he noted on 20 February, 'two Englishmen arrived on the island, and as soon as they landed they asked to see the Emperor . . . They handed over two packets . . . the next day he saw the foreigners again and they left immediately afterwards.'

On this occasion the Oil Merchant was reporting hearsay that was almost a fortnight old, so that there is no means of knowing whether the two men were anything but normal tourists and whether they were in Elba before or after the vital date on which Pons was first given his instructions. The Oil Merchant, moreover, had a different explanation for the comings and goings of mysterious Englishmen; he was convinced that the English were officially colluding in Napoleon's projected escape. On 17 February, in fact, he noted (wrongly) that Campbell had spoken at length with Napoleon before sailing for Livorno, and that Adye was supposed to have brought in a substantial sum of money for the Emperor. On 22 February he went even further, passing on a rumour which was soon to run across Europe and greatly embarrass both Campbell and his government. A staff officer, he said, had told him that the English had agreed that Napoleon should leave Elba for France, because they had made unsatisfied demands on Louis XVIII and Napoleon had promised to accept the British terms if he were allowed to return to Paris.

The Oil Merchant was sadly deceived by his suspicions. 'The keystone of my policy is the preservation of the Bourbons on the throne,' Lord Liverpool wrote to Castlereagh on 20 February; 'all other dangers may be regarded as contemptible when compared to those which would arise out of another revolution in France.' And though the government in London was keen to get rid of Murat as soon as possible the Prime Minister had already told Wellington on 11 January that military operations for this purpose were 'an absolute impossibility' in the present state of public opinion and finances. The only shred of truth in the Oil Merchant's suspicion lay in Castlereagh's belief that it would be better to shift both Murat and Napoleon by agreement rather than by force, and that this might be facilitated by 'a liberal proposition' in the form of cash and territorial compensation. After twenty exhausting and ruinously expensive years of war, and with a comprehensive peace now in sight at Vienna, the last thing that Liverpool and Castlereagh could have wanted was to see Napoleon at large again and making a new bid for power in Europe.

Yet the ambiguity of Campbell's position, which had worried him for months, was about to lead to a final and fateful misunderstanding. The Oil Merchant had a clear idea what was happening in the last weeks of February, but there had been so many false alarms that he could scarcely expect anyone on the mainland to take his latest news very seriously and in any case, as the pace of events accelerated, he could not deliver his warnings to anyone with the power to act in time. The only persons who might have done something effective were Campbell and Adye. If Mariotti's agent had been able to trust them and make contact with them, the story of the next two weeks might have been very different.

★

On 14 February, just before Campbell left and the Oil Merchant returned, everything seemed quiet on Elba. 'Napoleon remains shut up in his apartment without seeing anybody', Mariotti wrote on that day to Paris, 'and people say that he is afraid of assassins sent from France and Italy.' Yet that was a critical date, for it marks a decisive change of pace which was undoubtedly due to the appearance of an unexpected emissary from France. 'A person calling himself Pietro St Ernest has arrived here in the guise of a sailor from the bay of Spezia', Campbell noted on the following day; and on 18 February, when the Oil Merchant paid a courtesy call on Madame Colombani he also learned about the stranger. 'A few days ago,' she told him, 'a distinguished personage disguised as a sailor came from Lerici in a felucca', and she added that the man had left soon after some secret conversations with Napoleon which had evidently raised the Emperor's spirits. There had been so many inconclusive reports of this kind that both Campbell and the Oil Merchant had come to note them without much excitement, but for once the palace gossip was true: this visitor was really noteworthy.

He was Fleury de Chaboulon, an unemployed official of thirty-six who had been a sub-prefect in Burgundy during the last months of the Empire and earned the Legion of Honour for his vigorous part in the defence of Reims. Before he set out on his difficult journey to Elba he had been to see Maret, who had been Napoleon's foreign minister and who had a long experience of confidential diplomacy; and he seems to have brought useful information at a critical moment, even if his vanity and his romantic taste for conspiracy later led him to exaggerate his own role in events. 'With a word from me Napoleon would have been lost', he wrote grandiloquently in the memoirs he published four years later to prove that he had been the arbiter of destiny. 'With a word I could have saved Louis.' And to underline that self-important thought he gave Napoleon the most unlikely line of all. 'Had it not been for you,' he made Napoleon say, 'I should not have known that the hour had struck.' It is not surprising that Napoleon scrawled sarcastic comments all over the copy of Fleury's book which eventually reached him at St Helena.

Fleury was vague about dates. Pons said that he left Paris before Christmas, and he claimed that he had run into all kinds of trouble on his devious journey to Elba, including a fever in La Spezia and an adventurous crossing from Lerici with a gang of smugglers. He had certainly taken at least a month on the way, and much of the news he brought was stale by the time he arrived. All the same

Napoleon listened closely while he went over the familiar catalogue of woes and confirmed the reports of earlier travellers from France. The country was in a sorry and unsettled state, saddled with debt from the wars and smarting under a humiliating peace. Taxes were high, property values were depressed by the threat of restitution which hung over those who had bought expropriated land, and the priests were trying to restore the religious practices and the unpopular tithes of the old regime. Manufacturers were being driven out of business by the ending of war production and by foreign competition; tens of thousands of men were out of work; and there had been food riots in some of the ports against the export of grain to England. Everyone seemed to have a grievance, and political life was degenerating into a round of complaints and recriminations and half-hearted conspiracies. The liberals who had hoped for a new constitutional system were disillusioned, the old revolutionaries were reviving the Jacobin war-cries and thinking of a new appeal to the faubourgs, and—most ominously for the Bourbons—both the men who had been retained in the army and the mass of discharged soldiers were thoroughly disaffected, sighing for past glories and longing for the Emperor's return. Even the ministers knew that the government was close to the point of collapse. 'We are really going on very badly', Jaucourt wrote to Talleyrand on 25 January, 'and we must do better if we do not wish to perish completely.'

Napoleon was already aware that such an improvement was improbable, and that if he chose his moment well he might return to France and chase the Bourbons out of the country with very little trouble. But the timing was vital, and he was aware that he could all too easily miss his chance. He might land too soon, while Louis could rally enough support to defeat him, or at least to make him fight the civil war which had been avoided by his abdication in April 1814. If he waited too long, however, he might be anticipated, for Paris was buzzing with intrigues as the politicians sought for a way to get rid of the 'Fat Pig', as the soldiers called Louis, without having to put the Corsican Ogre in his place. There was talk of another republic, for instance, of grumbling among the ambitious marshals who resented the snubs they and their wives had suffered at the Bourbon court, of a faction which thought of calling back Marshal Bernadotte, who had left Napoleon's service to become Crown Prince of Sweden. Even if some of Napoleon's own supporters in the army were to stage a successful revolt, it was not quite certain that they would call the Emperor back; and if they did there were all the dangers of delay, confusion, and foreign intervention while he was being fetched from Elba. From a phrase that Napoleon once let drop, indeed, it seems that he had received and rejected one approach of this kind before the end of 1814. It would be far safer, and far more certain, to take matters into his own hands, and rely on no one else.

In this confusing situuation Napoleon was certainly glad to have direct word from the trusted Maret, who was in close touch with Marshal Davout, Thibaudeau, Lavalette, and other subordinates who had served him loyally and never taken the King's hand or money; with Hortense, now called the Duchess of St Leu, who ran a kind of Bonapartist salon with some other ladies who regretted the loss of the influence they had enjoyed in the days of the Empire; and even with Fouché, whose attempt to manipulate the situation and rally the Bonapartists in France without conjuring up Bonaparte himself, was undoubtedly a factor in prompting Napoleon to take the initiative. For Maret, through Fleury, could tell Napoleon what his friends were thinking without saying anything that might compromise them too much if Fleury were to be arrested, or otherwise led into indiscretion. He could say, for it was common knowledge, that much of the army counted the days until the Emperor returned. There was no great danger in that, because police agents were reporting the same rebellious mood from every garrison town in France. And he could also pass on the gossip about other people's plans and plots, for there was no risk to the Bonapartist faction in telling tales about its rivals.

It seems that some such piece of gossip was the most important news Fleury carried to Elba. Maret was no doubt anxious about some of the desperate-tempered young generals who tried to involve both Davout and himself in a scheme for a military rising—men such as General Flahaut, the illegitimate son of Talleyrand and the lover of Queen Hortense, General Drouet d'Erlon, General Sebastiani, and General Excelmans. There had been whispers all autumn about a military plot. 'The close surveillance that I maintain over the discontented generals shows that they visit each other a good deal,' Beugnot reported to Louis on 25 November; 'they mutter; they fear to see themselves supplanted by rivals favoured by the Court; but so far I can find no sign of combination nor of any attempts to rouse the people of the faubourgs.' It is clear that the director-general of police was not inclined to take the dissidents seriously at that time: 'amidst so much exaggeration, irrational

fears, and accusations inspired by dislike and party spirit', he told
Louis next day, 'I can nowhere see any kind of preparation, no
plan, no system, no chiefs, no stock of arms, no complicity with the
Army or the National Guard'. Even though the military conspiracy
began to take a more definite form towards the end of the year
Fleury could not have known anything significant about it. For one
thing, Maret would have been most reluctant to confide in a stranger, even if he had good credentials: Napoleon's intimates were all
trained to avoid careless talk. For another, Fleury must already
have left before the conspirators had decided what to do.

That seems quite clear from the recollections of Lavalette, one
of the most sensible and reliable of all the Bonapartists in Paris.
Sometime in the New Year, he said, he was approached by General
Lallemand, one of two brothers who were among the leaders of
the conspiracy. 'He wished me to take an active part in the plot',
Lavalette recalled, 'and especially that I should undertake the duty
of bringing it to the Emperor's knowledge.' There would have been
no need for that request if Fleury had already been entrusted with
the task. Lavalette gave a cautious and critical reply. He told Lallemand that the plotters were taking 'a great liberty' in involving the
Emperor without his consent; that they might be endangering him,
or at least providing the Allies with an excuse to send him 'to the
other side of the world' if they were caught; and that in any case
Napoleon might well have plans of his own, unknown to anyone in
Paris, which their scheme could 'impede or ruin'. When Lallemand
replied, brusquely, 'things have gone too far to delay', Lavalette
hurried round to see Maret for himself. And Maret then said explicitly that Fleury, who had left 'more than a fortnight ago', could
not have carried any warning to Napoleon. 'At the time of his
departure', he confided to Lavalette, 'the soldiers' plot was not yet in
existence, or at any rate I did not know about it.'

For the same reason it is unlikely that Fleury knew anything
about the talks which Fouché had with Maret and other Bonapartists,
for these discussions, too, apparently took place in late January or
early February, when Fleury was already well on his slow way to
Elba; and they seem to have been linked in part to the military plot
(for that was the context in which Maret reported them to Lavalette),
and in part, according to Fouché's own account, to his
favourite notion of setting up Marie-Louise as regent for the infant
King of Rome. But it is possible that Maret told Fleury about some
of Fouché's other machinations. For that tireless opportunist was

also trying to ingratiate himself with the royalists, and the constitutionalists; he was secretly writing to Metternich; and he was undoubtedly aware of the proposal, canvassed by a few of the more disgruntled marshals, to put the duc d'Orléans on the throne instead of Louis. And whether Fleury got his information about that intrigue from Maret, or Davout, or picked it up from other sources, it was almost certainly that news which upset Napoleon and persuaded him that he must move as soon as possible.

It was not, in fact, a very serious plot, for the duke was too cagey to be drawn into an affair that was little more than a grumbler's tea-party, but neither Fleury nor Napoleon were in a position to know that and, so far as Napoleon was concerned, it would be unutterably galling to gamble everything on a return to France and then to find that it was all for nothing—to discover that the son of the old Bourbon roué and revolutionary they called Philippe Égalité was already installed as King of France, and that the springs of discontent that might have driven a Bonapartist rising were all unwound. Napoleon made that point himself. 'Fleury de Chaboulon', he wrote years later at St Helena, 'brought me news at Elba of the conspiracy in favour of the duc d'Orléans.' The ensuing reference to Davout seems odd in view of Maret's insistence that Fleury was not the bearer of any request for Napoleon to go back to France. That, Maret said, was a matter for the Emperor himself to decide. But Napoleon seemed to remember differently. 'Davout', he added, 'was particularly urgent for my immediate return. He was quite right, for the coronation of the duc d'Orléans would have been for many persons, and especially for the foreign Powers, a sort of compromise between the Revolution and the Restoration.'

All the evidence thus suggests that Napoleon was already preparing to leave when Fleury reached Elba, and that the news Fleury brought merely made him more eager to go before he was forestalled. He was never a man to confide in his subordinates, and on this occasion he needed more to surprise his enemies than to warn his friends. John Cam Hobhouse, who went to Paris as soon as Napoleon returned, and spoke to many persons close to the Emperor, came to the conclusion that 'there was no corresponding scheme laid at Paris for the restoration'. Writing a letter to England on 27 April he said that 'the whole prospect and execution are to be attributed solely to the daring determination of Napoleon himself to recover his crown, most happily coinciding with the actual conditions and general feeling of France', and he noted several fragments of gossip that appeared to confirm that opinion. One informant said that the Emperor had declared that his only merit lay in making 'a good guess as to the actual situation'; another reported that he had told Fouché 'that the revolutionary spirit in France, being brought prematurely to a head, obliged him to take advantage of the general feeling three months sooner than he would have wished, and whilst the armies of the Allies were still in a position to recommence operations against France'.

★

Through all the ups and downs of life in his island kingdom Napoleon could never have abandoned the hope that he would somehow make his way back to Paris. There had been depressing weeks, when such a reversal of fortune seemed unlikely, or impossible; when he had thought that his last chance of useful employment might be the command of an Austrian army, defending Vienna against the Cossacks; or when the only prospect of a change was transportation to a more distant place of exile. Yet he had always been attracted by the idea of revenging himself upon the Bourbons, and expunging the shame of his flight across France. 'In any case,' he was to say afterwards, 'my return from Elba shows that I am not a nincompoop.' What stung him, he declared on another occasion, 'was the accusation of cowardice; in all the slanders it was said that I feared death, that I had never run any personal risk. At last I could stand it no longer.'

A sudden descent on France, moreover, was quite in keeping with Napoleon's manner of waging war. As he reminded Colonel Gourgaud on St Helena, when he was looking back over his career as a soldier, he had won so many battles by simple means rather than by subtlety, by gambling on surprise, by swift movements which struck at the main force of the enemy and crushed it while the defenders were still reeling from the shock. In the conditions of 1815 the military and political imperatives combined to make a direct march on Paris the only sensible policy.

That should have been obvious to anyone who had studied Napoleon's tactics and given any serious thought to the situation in France, but even men as close to events as Campbell and Mariotti shared the widespread conviction that if he did escape from Elba he would either go to meet Murat in Naples or engage in what Campbell quaintly called 'some eccentricity' elsewhere in Italy. This obsession with Italy was based on two assumptions: that Napoleon needed Murat because he had no other potential ally, and that it would be easy for him to land in Tuscany and lead an Italian insurrection against the Austrians. Both were false.

The first error was to think that Napoleon had anything to gain by joining forces with his unstable brother-in-law at the bottom of the Italian peninsula, far from the centres of power and the classic battlefields of Europe. On the contrary, he was well aware that such a flight would confirm every suspicion of collusion and lead at once to war with Austria, France, and possibly England. The second error was to see Napoleon as an enthusiast for the liberation of Italy when his whole career showed that he would cynically encourage or betray the Italian patriots as it suited him. All his covert links with them, and his open interest in Italian affairs, amounted to little more than a calculated diversion, which generally distracted attention from his continuing attachment to France and was to serve at least one particular purpose when it sent Campbell off on the wrong track at the critical moment.

Napoleon himself was quite frank about the futility of an Italian adventure. Writing at St Helena he claimed that he had rebuffed several revolutionary proposals from Italy. 'You can do nothing without France', he supposedly told the nationalist emissaries, 'for you will be wiped out by the Austrians; you have neither the makings of an army, nor enough arms, nor a single fortress. . . . If you love your country, remain cool and collected, let things take their course in France; wait until there are changes in France itself.' Even if the wording was apocryphal the idea was sensible, and it explains why Napoleon never really considered the leadership of an Italian revolution.

He did not even consider Italy to be a useful stepping-stone to other destinations. Once Napoleon landed in France he would be back on his own soil, with his own people, and a good chance that his Guard battalion could be quickly reinforced as it met well-trained regiments ripe for desertion. Within a few days he could have the makings of a real army, and within a few weeks, with luck, he could be in charge of his Empire again. But Italy was a disorganized country occupied by Austrian troops, and Napoleon knew that they could soon round up his little band of men and crush a revolution in Milan and Turin as easily as he had done years before. He had no hope at all of quickly raising an army capable of surviving, let alone of striking towards Vienna, as he had done in 1796, or of marching over the snowbound Alps into Savoy and on to Paris.

Napoleon had made his reputation in Italy, almost twenty years before. The only thing he could do there now was to lose it. For all the risks, Provence was the only possible place for a landing.

17

THE LAST OF ELBA

Campbell was due to leave Elba on 16 February for what he called another 'short excursion to the Continent for my health', and on this occasion Napoleon left the valedictory interview to Bertrand. It was a curious meeting which left Campbell puzzled, for Bertrand spent most of the time arguing about the status of Ricci, who had lately been used by Campbell as something between an informer and a shipping-agent. 'I cannot precisely account for this sudden and apparently useless stir about Mr Ricci's powers,' Campbell wrote uneasily; 'it might be to intimidate him from giving me any information of what passes on the island.' This was a sobering thought. As he proposed to rely upon Ricci to let him know 'if anything extraordinary occurred' during his absence, Campbell naturally wondered whether to take this sudden interest in Ricci's activities as 'a proof of some improper and guilty connection' which Napoleon and Bertrand were anxious to conceal. He also thought that the discussion showed 'a want of delicacy and politeness' which 'arose from a wish to disgust me, and induce me not to remain'.

Once again Campbell felt that he was 'remaining on Elba as an obnoxious person, upon a kind of sufferance', but he braced himself to his duty: 'however disagreeable the prolongation of my stay might be under such circumstances', he noted in his journal, 'I resolved to remain, being in daily expectation of the Congress terminating'. These dutiful intentions, however, were for the future, and though Campbell put his suspicions into an anxious despatch to Castlereagh on 15 February, he did not let his feelings change his immediate plans. After a 'loud and warm' exchange with Bertrand about Ricci, and about the recent building of a gun platform on the Palmaiola rock in the Piombino Channel, he prepared to leave for Livorno on his way to Florence.

That was exactly what Napoleon wanted. The sooner Campbell left Elba and the longer he stayed away the better, for it was risky to begin his final preparations while Campbell was free to wander about the island and the *Partridge* was in a position to stop any ship from leaving it. Napoleon, of course, could arrest Campbell and threaten or sink the *Partridge* with the fortress guns. But that would have been a remedy of last resort, for the most brilliant feature of Napoleon's plan for escape was his notion that he might manage the whole thing peacefully. 'Don't worry,' he told Drouot before they left. 'We are going to reach Paris without a shot being fired.' Campbell's convenient journey to Florence thus provided the final date in Napoleon's provisional timetable. Since it would take at least ten days the Emperor would have to leave Elba on or before the night of 26 February 1815.

Napoleon sent the little *Étoile* to follow the *Partridge* out of port and make sure that Campbell had really gone to Livorno, and then he wrote a most revealing order to Drouot. The *Inconstant* was to be brought into the harbour at once and 'painted like an English merchant brig'. The vessel was then to be loaded with enough victuals to keep 120 men for three months (or, if the figures are re-calculated, to keep 1,000 men for about ten days), to be furnished with as many small boats as possible, and to be 'in all ways ready for sea' by 24 or 25 February.

A few days afterwards, when Napoleon told Drouot explicitly that they would soon be leaving for France—'the whole of France regrets me and wants me back'—the sober but loyal general reacted strongly. 'Struck with astonishment, I expressed my opposition,' he said a year later, when he was court-martialled for his part in the escape, 'but I was bound to Napoleon by my oath.' Like Bertrand, who was similarly worried by the prospect of a new military adventure which might end in the further exile or death of all of them, Drouot had a strict sense of his duty, and for all his misgivings he would do his best to ensure that the Emperor's plans were successful. On 22 February he told the saddler to prepare his campaign riding bags and map-case. On the same day Napoleon called on Peyrusse to tell him to prepare to leave. He was to load the gold reserve into travelling-cases, and to cover the money with books from the Mulini library. 'You can get your baggage packed as well', Napoleon said cheerfully and almost casually, as if the flight from Elba was to be nothing more than a straightforward journey to France. But when Peyrusse hastened to discuss this cryptic conversation with Drouot he found the general 'in a solemn and reflective mood'.

Pons, Peyrusse, and Drouot were senior officials who had to be trusted with the secret now that the decision was finally made, but to prevent other people jumping too easily to the right conclusion

Napoleon kept up a flow of distracting instructions about Elban business. On 16 February he gave away parcels of land for his soldiers to use as allotments, and urged them to get on with their spring sowing; at his mother's suggestion he set men planting mulberry trees along the road to San Martino; the painters were put to work in the Villa Mulini; and Bertrand was instructed to start making the arrangements for Napoleon, Letizia, and Pauline to spend the hottest part of the summer at Marciana. The Oil Merchant, just returned to Portoferraio, said that the men were working on their allotments 'more as if they had just arrived than as if they were getting ready to leave', but he at least was not deceived.

On 19 February, when he reported the arrival from Marseilles of a merchant named Charles Albert whom he suspected of being a Bonapartist messenger, he quoted the man as saying that 'the sight of Napoleon's hat on a French beach would be enough to rally every Frenchman to his side'; and over the next few days he saw ample evidence that Albert's prediction might soon be tested. On 21 February he noted that stores were being moved and packed, and that an artillery park had been set up near the port. On 22 February he heard that three companies of the Guard had been kitted out with new topcoats and two pairs of boots for each man, that the horses of the Polish lancers were being brought back from pasturage on Pianosa, and that sixty crates of cartridges and other supplies had been carried on the *Inconstant* during the night.

Ricci was also becoming alarmed by the buzz of military activity. He was not such a cool and accomplished spy as the Oil Merchant, but he had seen enough to make him send Campbell a letter on 18 February which 'contained matters of such nature as to excite the gravest suspicion'. Like the Oil Merchant he had picked up the barrack-room gossip. 'The troops', he told Campbell, 'are full of expectation of some great event.' He too had heard that the *Inconstant* was taking on 'military stores'; he was correctly informed that Pons had sent two empty ore-carrying feluccas from Rio to Portoferraio, and was trying to charter or buy at least one more merchant ship; and he had picked up the disturbing news that Colonna had again passed through Longone on his way to Naples.

Events were now moving fast. Colonna was carrying an 'important and pressing' communication for Murat. 'I beg you to believe all that he tells you,' Napoleon wrote to his brother-in-law. 'He is authorized to sign any convention that Your Majesty might desire relative to our affairs.' Colonna apparently delayed his crossing until he heard that Napoleon had definitely sailed from Portoferraio, for he did not arrive in Naples until the first days of March. As he had been sent as an ambassador to restore the formal contacts which had been broken when Murat had defected to the Allies, there was no point to his mission unless he could say that Napoleon was already on his way to Paris to reclaim his throne. For he was to tell Murat that the Emperor had forgiven him for 'all his mistakes'; to insist that Murat should be careful, since Napoleon was only too conscious that the news of his escape might provoke King Joachim into some act of folly; and to ask Murat to do all he could to neutralize the Austrians—either to persuade them that Napoleon wanted peace or, if he failed to convince them of that, to keep the Neapolitan army in a state of alert and thus deter Francis from sending troops from Italy to help the Bourbons resist Napoleon in France.

It was a sensible message, but it was not delivered to a sensible man. The King of Naples, Napoleon wrote later, 'seemed astonished by so much moderation', and at the same time as he impulsively offered Colonna ships, troops, and money to support the Emperor's expedition he began to make wild and ultimately fatal plans to rouse Italy on his own behalf. Napoleon had not wanted anything so dramatic or so compromising. His only specific request for military help from Murat was carried by Pietro Santa, nominally a member of Pauline's household and actually a confidential courier, who overtook Colonna on the way. Santa's mission was secret but straightforward. He was simply to ask Murat to send his naval squadron to cruise in Elban waters during the critical days at the end of February and the beginning of March, as it had done when there seemed to be a threat from the Barbary pirates. Napoleon did not want to be carried or escorted to France by Murat's men-of-war. He was determined, he said, to return in a French ship flying the French tricolour. But if he fell foul of the royalist frigates or the *Partridge*, or had to turn and run for it, two 74-gun ships of the line could provide a valuable diversion or cover for his retreat to Naples. There would be nowhere else to go.

Murat hesitated, while Caroline argued that he would be foolish to risk the throne of Naples by supporting her brother's escape; and when the Neapolitan ships were sighted off Elba early in March they were too late to make any difference. The Emperor had already managed without them.

★

This later engraving of Napoleon's departure from Elba on 26 February 1815 accurately portrays the *Inconstant* and the other vessels involved

Of many pictures showing Napoleon's meeting with the royalist troops at Laffrey, this is the one which most accurately portrays the setting and the units involved

'The Congress Dissolved Before The Cake Was Cut Up'. Caricature by George Cruikshank. 6 April 1815

'One can't take a step without being noticed', the Oil Merchant complained on Thursday 23 February. There were now so many guards and police agents patrolling Portoferraio that he decided he could safely do no more, and that if he wanted to reach the mainland in time he would be wise to ask for his papers and leave next morning. 'Today everyone is speaking quite plainly about a departure,' he wrote. 'I begin to be convinced that it is probable.' In the course of the day, moreover, the Emperor went down to the harbour to inspect the *Inconstant* and the *Étoile*, which were being loaded with food and water before moving out into the roadstead; and he also went on board the *Saint Esprit*, a merchant ship of about 200 tons which had come in from Genoa and then been requisitioned to form part of the flotilla which was soon to sail for France. With the *Caroline*, the two feluccas which Pons was to send round from Rio, and an Elban ship called the *Saint Joseph* which had been chartered from a local merchant, Napoleon would have only seven slow ships for his expedition. In naval terms the enterprise was foolhardy if not farcical.

Late that night the whole scheme seemed suddenly to be at risk, for the *Partridge* unexpectedly came in under a bright moon and anchored only a few cables away from the *Inconstant*. Napoleon assumed that Captain Adye had no particular reason for suspicion, or he would not have brought his sloop up under the guns of the forts, but there was a danger that he might ask why the *Inconstant* had been painted like a British brig and why there was so much activity in the harbour. Before dawn, therefore, Napoleon told Captain Chautard to take the *Inconstant* to sea until the *Partridge* had sailed again. He suspended the loading of munitions and stores, and he again set squads of soldiers to garden the plots by their barracks. Adye went ashore about nine in the morning of 24 February, accompanied by half a dozen English tourists he had brought to the island. After he had been told by the port captain that the *Inconstant* was off to Livorno for repairs he walked up to the Villa Mulini to get Bertrand's assurance 'that Napoleon Bonaparte was still on the island', watched the guardsmen 'all busy carrying earth and planting trees', and sailed soon after two o'clock to look at the new fortification on Palmaiola which had been the cause of Campbell's parting complaints. Napoleon could relax again.

Adye must have been a naive and surprisingly unobservant naval officer, for the anxious Campbell had sent him back to Elba specifically to find out if Napoleon was 'still on the island'. He was under general orders from the Admiralty to interfere if he found the Emperor in the act of escape with evident warlike intentions. He had spent several hours in a small port where men and ships were being prepared for an early departure, yet he had seen nothing to worry him. The only sign of suspicion in his report was a note that later in the afternoon of 24 February he had been 'rather surprised' to see the *Inconstant* returning to Portoferraio in company with the *Étoile* and the *Caroline*—the smaller vessels having been sent to fetch the brig once Adye had been successfully deceived. Luck, as so often, ran Napoleon's way. Ricci did not manage to get a message to Adye while he was ashore, and the Oil Merchant was so convinced that the British were colluding with Napoleon that he let Adye come and go without any attempt to reach the *Partridge*. He instead spent a futile day trying to get permission to sail in a fishing-boat. 'I prayed, I offered bribes, I made promises,' he noted, 'but all to no effect.' That night the loading went on steadily.

Yet the *Partridge* was still a threat. It met the two feluccas that Pons was bringing round from Rio, and gave him a fright, though with presence of mind he sent over an invitation to Campbell to dine next week, which seems to have satisfied Adye that all was well. And then the sloop lay in the Piombino Channel all night, waiting for a breeze to carry it up to Palmaiola. Later on Saturday, after Adye had been refused permission to land on the fortified rock, he sailed back past the entrance to Portoferraio on his way northwards. Once again the gunners ran to their posts in the forts and there was a general alarm. Once again Adye missed his chance of surprising the Emperor on the point of departure. He simply reported that at six o'clock he had plainly seen the masts of the *Inconstant* in the harbour as he held his course for Livorno. With such light winds he was worried about keeping his appointment to meet Campbell there on the following afternoon.

Saturday had been a busy time on Elba. Although Napoleon himself had so far said nothing, everybody seemed to know that the troops would soon be leaving, and there was a flurry of last-minute preparations, carousing, and farewells. 'I seem to have been transported into a Tower of Babel', the Oil Merchant remarked, and he was bitterly frustrated because he could find no means of leaving it. For the police had taken back his passport, there were military patrols everywhere, and there was such a strict ban on sailings from the island that even the fishermen could do no more than ply about the harbour round the loading flotilla. He tried to bribe several boatmen and found that 'they refuse at any price to take a passenger over to the mainland'.

Napoleon gave out that he was indisposed and he stayed in the Villa Mulini all day. There was much to do. He had to draft the three proclamations he proposed to distribute when he arrived in France, for Broglia the printer was standing by to run them off that night. He had to settle the loading arrangements for his ships with Drouot and Bertrand, and to tell them for the first time in detail what he proposed to do when he landed in Provence. And when all the work was finished he proposed to spend a quiet evening with his mother and his sister, dining early and playing cards as they had done so often in the course of the winter.

This was clearly an emotional occasion, for Letizia and Pauline were the only two people in his family to whom he was deeply attached. But like all the dramatic moments in the Napoleonic legend it was later embellished to theatrical effect. Letizia described how Napoleon went out to pace the terrace in the waning moonlight, how she followed him and learnt that he was leaving for France next evening, how he had asked for her opinion, and how, with a sense of fatality, she had wished him a hero's fate. 'If Heaven intends that you shall die, my son, and has spared you in this time of ignominious exile, I hope you will not perish by poison but with your sword in your hand!' It was a touching scene, as was Pauline's farewell gift of her treasured diamonds to swell Napoleon's war chest—the diamonds which were found by the Prussian cavalrymen who looted his carriage after Waterloo. Even in these last domestic moments, however, Napoleon seems to have played out his role as a man sacrificing everything to his destiny and aware that Clio as well as Melpomene was watching him. The muses of history and tragedy were never far from his thoughts.

In Florence that week the anxious Campbell had been given an equivocal reception. He had gone to hand in his latest despatch for Castlereagh and to share his growing uneasiness with Lord Burghersh, but instead of sympathy or support he had been greeted with a reprimand for 'the improper manner' in which he had lately been carrying out his duties, and begged to remain 'more constantly at his post'. Burghersh spoke bluntly to Campbell, and then formally complained about him to Castlereagh in a letter he sent on 3 March. 'His absences from Elba', Burghersh said, 'have been constant and at times of considerable duration.' Yet Campbell was also reassured by his conversations with Edward Cooke, an under-secretary from the

Foreign Office with considerable experience of confidential missions, who had been with Castlereagh in Vienna and had just come down to Italy on leave. 'You may tell Bonaparte that everything is amicably settled at Vienna,' Cooke replied complacently when Campbell asked him whether there was any news about Napoleon's pension or of his wife and child; 'that he has no chance; that the Sovereigns will not quarrel. *Nobody thinks of him at all. He is quite forgotten—as much as if he never existed!*' It was not irony but relief that led Campbell to underline the last two sentences when he entered them in his journal. They made him feel, he said, that his 'near view' of Napoleon had led him to exaggerate the danger. As soon as Campbell returned to Livorno on Saturday 25 February, however, he was alarmed by the news that had somehow reached Mariotti. It was now evident, he noted, 'that Napoleon was on the point of embarking a military force with stores and provisions', and he immediately sent fresh despatches to Burghersh and Castlereagh. He assumed that Murat and Napoleon had both learnt 'the decision of the Congress' and decided that the Powers would soon dethrone the King of Naples. 'I think', he wrote, 'it is almost certain that Napoleon is prepared to join Murat, in the event of the latter throwing down the gauntlet in defiance of the sovereigns of Europe.' It was a reasonable, disturbing, and quite misleading conclusion.

★

The morning of Sunday 26 February was calm and clear, with a light breeze. Napoleon was up at six, and he took unusual care in dressing himself for the day, putting on white silk stockings, white kersey-mere knee-breeches and waistcoat, a black silk stock, and the dark blue coat with white lapels and scarlet cuffs worn by grenadier officers in the Guard. He fastened on the sword he had carried at Austerlitz, pinned the Legion of Honour and the Iron Crown of Italy above his heart, and made sure that his valet Marchand had a tricolour cockade ready to replace the Elban emblem in his celebrated black beaver hat. After ten months of exile the Emperor was about to come into his own again.

He certainly gave that regal impression when he appeared at his customary Sunday levée, which was crowded with Elban notabilities and his own retinue. Like the townspeople who had climbed up the steep alleys to stand outside the Villa Mulini and shout 'Vive l'Empereur!', they had already guessed that this would be the last of such occasions. Rumour had run fast in Portoferraio in the past few days. But Napoleon was cautious to the end, and though he went on from his usual opening pleasantries to announce that he would soon be leaving Elba he had nothing to gain and much to lose by casually revealing his destination to this collection of palace gossips, or to any of those to whom he said goodbye in the course of the day. Pons, who was an eye-witness, said explicitly that the Emperor 'covered his plans with a veil of mystery' at the levée; and all the reports of speeches about the miseries of France and the crimes of the Bourbons seem to be subsequent improvements upon the facts. Secrecy was Napoleon's normal military practice. For several days his troops had been speculating like gamblers before a big race, each man confident of knowing the name of the winner. Yet they sailed that night without knowing for sure whether they were going to Naples, Genoa, Toulon, or Marseilles, or even to Alexandria or Tunis.

Each hour now had its duties and excitements. At nine Napoleon went to Mass in the town church. At ten he walked out into the town square to review the National Guard before the men marched off to take the place of the regulars who had previously guarded the port and the fortresses. And at eleven a small boat ran into the beach below the Villa Mulini with the news that the coast was clear. At noon the drums beat out the call for assembly, and as the soldiers returned to their barracks they were told to kit themselves out in battledress and be ready to leave at four o'clock.

The Oil Merchant was almost carried off with them. Early that morning he had at last persuaded a boatman to attempt a crossing to Piombino, although the man had demanded the exorbitant sum of sixty livres for this risky enterprise. But they did not get far. As they tried to slip past the *Inconstant*, anchored at the mouth of the roadstead, they were challenged and told to turn on pain of arrest. He then rode over to Rio in the hope of finding a subornable boat-man at the smaller port. This time he was watched so closely that he thought it wise to go back to Portoferraio, where he arrived just before the town gates were closed and the troops began to embark. On the quayside he met Cambronne, who told him brusquely that as a former officer he should sail with the flotilla and that there was a place for him on one of the smaller vessels among the Elban volunteers. 'Faced by this unexpected order', the Oil Merchant said, 'I was hard put to find a reason why I could not leave.' It was only by appealing to his old comrade Colombani, and by promising that he would follow the Emperor as soon as he had settled his business affairs, that he managed to avoid immediate enlistment in the adventure which he had predicted and been unable to prevent.

★

There was a good deal of haste and confusion in Portoferraio that afternoon, as crowds gathered to watch the ships being readied for the voyage. Napoleon himself was dealing with last-minute business. He spent some time burning papers, and after he left there were scraps of torn and crumpled paper on the floor, including drafts of his memoirs. He went out on a final inspection. He sent for Pons, who had hoped to be made governor of Elba, and told him that he would be leaving that night without any chance to say goodbye to his family in Rio. He also sent for Dr Lapi, the commandant of the National Guard, promoted him to general, gave him Drouot's post of governor, and instructed him to go on ruling Elba with a junta of local worthies. The King of Elba was leaving his subjects to their own devices as suddenly as he had first come among them, and soon after nightfall he and all the men who had joined him in his temporary kingdom would be gone. By seven in the evening, after a quiet dinner with his mother and his sister, and a tearful farewell from the women of the imperial household, there was nothing left to do but to put on his famous grey travelling coat and to get into Pauline's tiny pony-cart and ride down to the Water Gate. Bertrand rode beside him, and behind walked Drouot and Peyrusse, Pons, and the other members of his suite who were going with him.

Cambronne had already got his men on board the flotilla. With bands playing and flags flying they had marched down through the twilight, past the windows illuminated in Napoleon's honour, under the coloured lanterns in the square, and through a crowd which called out and cried in its excitement. The grenadiers of the Guard had come first, for they were to go on the *Inconstant* with the Emperor and his personal staff. That accounted for almost 500 men. The remainder of the Guard, the Polish lancers and the civil employees were loaded on to the *Saint Esprit*, the Corsican battalion, the gendarmes, the gunners, farriers, and other small groups of specialists were divided among the smaller vessels, and the *Caroline* was to carry the marines of the Guard. Altogether there were just over 1,100 men, 40 horses, Pauline's *berline* and 4 cannon. It was a small force to gamble against France, let alone the Powers of Europe.

The farewell ceremony was a mirror image of the welcome that Traditi and the other notables had managed at equally short notice ten months before. As the same little knot of men stood again on the jetty in front of the Water Gate the crowd fell silent, but no one could hear what Traditi was trying to say through his sobs. Then Napoleon spoke a few simple sentences of thanks. 'I shall always be grateful,' he said: 'you may count on my remembrances . . . Take care of my mother and my sister.' Pons was as moved as anyone. 'It was tears, tears, tears', he wrote afterwards.

They were singing the 'Marseillaise' as Napoleon went down the steps into the *Caroline*, which was to take him out to the *Inconstant*. There was so little wind that the men had to row all the way, and as the little craft passed each vessel in the flotilla there was a burst of cheering. It was the last of Elba. As the Emperor boarded the *Inconstant* a single cannon signalled that the escape had begun.

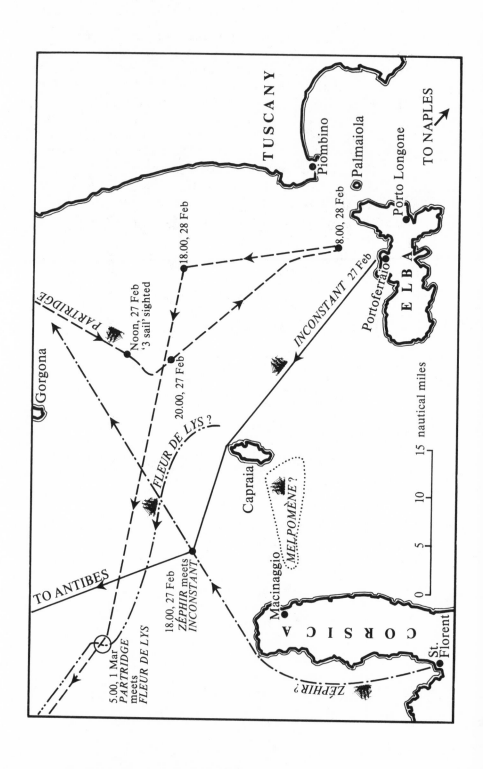

TUSCANY

Piombino

Palmaiola

Porto Longone

TO NAPLES

18.00, 28 Feb

Noon, 27 Feb
'3 sail' sighted

PARTRIDGE

8.00, 28 Feb

Gorgona

20.00, 27 Feb

INCONSTANT 27 Feb

Portoferraio

E L B A

FLEUR DE LYS ?

Capraia

MELPOMÈNE ?

0   5   10   15   nautical miles

TO ANTIBES

18.00, 27 Feb
ZÉPHIR meets
INCONSTANT

Macinaggio

C O R S I C A

St. Florent

ZÉPHIR?

5.00, 1 Mar
PARTRIDGE
meets
FLEUR DE LYS

18

THE FLIGHT OF THE EAGLE

It was nearly a fiasco. Once Napoleon decided to leave he had to take his chance with the weather. If there was no wind, or if there was a strong north-easter of the kind which was common at that time of year, he would never be able to work his ships out of the roadstead, and they would be at the mercy of the *Partridge* when the sloop came down the wind from Livorno. What he needed was a steady southerly breeze which would carry him well clear of Elba while it was still dark, partly because he did not want anyone watching from the island to know whether he was heading for France or Naples, and partly because he hoped to nurse his little convoy past the usual patrol line of the French frigates before the sun came up. And Sunday 26 February had promised well after an unusually calm week, for a suitable wind had blown down from the mountains and across the harbour all afternoon. But by the time the ships were ready to sail the wind had failed completely, and they lay motionless. The only way to get out of the lee of the island, to pick up the faint airs that were ruffling the open sea, was to row them out. It was midnight before they all cleared the lighthouse on the point at Portoferraio; at dawn on Monday they were still only six miles away, slowly heading west by north towards the blue hulk of Capraia, and the crowds which went up to get a good view from the forts could see the straggling line of sails for much of the afternoon.

Napoleon was apparently sailing into a trap. At noon on Monday, when his ships were close to Capraia, backing and filling in an effort to get round to the north of the island, the royalist frigate *Melpomène* was sighted to the south-west of it, in the passage between Capraia and Corsica. The second of the patrolling frigates was also in a position to intercept the convoy, for the *Fleur de Lys* was cruising somewhere to the north, between Capraia and Gorgona and close enough, her captain said, to see the *Melpomène* in the course of the day. Thus, by design or ill-fortune, there was a hostile frigate in each of the two channels which ran from Elba towards the coast of France. And beyond Capraia the French naval brig *Zéphir* was sailing towards Livorno on a course which would bring her right up to the *Inconstant* at dusk. Early on Monday afternoon, therefore, the *Inconstant* was at the south-east corner of a diamond which had Capraia at its centre and three French warships at the other corners. Even in such calm conditions they were far too close for comfort.

At this moment the trap was almost closed. A few minutes after the *Inconstant* claimed to have sighted the *Melpomène* the look-outs reported another strange sail coming down from Livorno. It was Adye in the *Partridge*, making painfully slow progress back to Portoferraio, and about to miss his chance for the third time.

★

Adye had reached Livorno at noon on Sunday. The wind had been so slack that he had taken eighteen hours to tack his way up from Elba, and as he arrived it failed so completely that the impatient Campbell had to watch the *Partridge* lying becalmed outside the harbour. It was eight in the evening before Campbell went on board, and it was nearly dawn before the *Partridge* picked up the light easterly breeze which was to carry her and the *Inconstant* towards Capraia that afternoon.

While Campbell waited through these lost and vital hours he read the disturbing report which Ricci had written on 18 February, questioned Adye 'with abrupt anxiety', talked at length to Mariotti, and wrote a long journal entry reviewing his conduct on Elba. He was already preparing his defence against the conflicting charges of negligence and collusion which were to plague him for the rest of his career.

In later weeks, when Napoleon's flight was debated in Parliament and the London newspapers, nothing more was said about the reprimand from Burghersh, for that would have undermined Campbell's claim that he had done all that his mission required of him. It would also have embarrassed Castlereagh, who was insisting that Campbell had been kept on Elba 'for the purpose of occasionally communicating with our government upon such matters as might pass under his observation' and that 'nothing more was ever contemplated'. He maintained that when the sovereigns despatched Napoleon to Elba 'it had never been in the contemplation of the parties that he should be a prisoner within that settlement', that 'the whole British navy would be inadequate' to prevent him leaving if he chose, and that Adye and other naval officers had 'an understanding with the

Admiralty, that if they suspected Bonaparte was contemplating a descent upon the opposite shores, they should immediately adopt such measures as would frustrate the attempt'. But the more these vague arrangements were discussed after the event the more it became clear that Campbell and Adye had been put in an impossible situation.

With hindsight it was clear that the Allies had been extraordinarily negligent in making Napoleon the King of Elba: as the Marquis of Wellesley put it on 12 April, there ought to have been 'due provision against his return to power'. And so Castlereagh and the Prime Minister kept shifting their ground. On 6 April Castlereagh told the House of Commons that there was 'an understanding with our officer stationed at Elba that Napoleon was to be confined within certain limits, and that he should not be allowed to exceed those limits'. On the next day, however, Lord Liverpool told the House of Lords that Napoleon 'was not considered in any degree as a prisoner on Elba'. With such equivocation at Westminster it is scarcely surprising that two subordinate officers were uncertain of their duty. Even if Campbell had been more conscientious, and Adye had been more observant, the 'understanding' under which they acted was so ambiguous that they could do nothing unless they actually caught Napoleon and his Guard at sea.

That was what Campbell now hoped to do. Anxious and angry, he drafted a firm declaration of intent. 'In case of Napoleon quitting Elba, and any of his vessels being discovered with troops on board, military stores or provisions, I shall request Captain Adye . . . to intercept, and, in case of their offering the slightest resistance, to destroy them.' And then he added two sentences that were clearly intended to be read whether Adye succeeded or failed. 'I am confident that both he and I will be justified by our sovereign, our country, and the world, in proceeding to any extremity upon our own responsibility in a case of so extraordinary a nature. I shall feel that in the execution of our duty, and with the military means I can procure, the lives of this restless man and his misguided associates and followers are not to be put in competition with the fate of thousands and the tranquillity of the world.'

Fifty miles away to the south, as Campbell wrote these fire-eating words, the first companies of the Guard were boarding the *Inconstant*, and Captain Chautard was looking as anxiously at her slack sails as Adye was whistling for a breath of air to carry the *Partridge* out of the Livorno roads. In the last hours the difference between escape and capture was to be little more than a few puffs of wind.

Napoleon was aware how close Adye came to running him down. When the look-out on the *Inconstant* saw the topsails of the *Partridge* at noon on Monday the two ships were about fifteen miles apart, and the gap was slowly closing as the English sloop picked up the variable airs. At two o'clock Napoleon decided that Adye had recognized him and that he would do better to seek help from the *Melpomène*, which he had sighted about the same time and away to leeward, than to risk a single-handed engagement with the *Partridge*.

This was an understandable decision, especially if Napoleon had been assured in advance that Captain Collet, commanding the French frigate, would ignore him if he was successfully passing to the north of Capraia, and would actually help him if he ran into serious trouble. 'She was about twelve miles away', Napoleon said afterwards in words that gave precisely that impression, 'and she was not bothering about our flotilla.' He then described how, as the *Partridge* came on, he ordered his smaller craft to make for the *Melpomène*. 'We knew enough about the feeling of the officers of these vessels, let alone the crews, to be sure that they would hoist the tricolour flag and defend the Emperor against the English ship.' An hour later, with the flotilla still caught in a patch of calm, these revealing precautions proved to be unnecessary. While the *Partridge* 'had come much nearer and could be clearly distinguished', Napoleon added, 'she did not seem to be concerned with us and she was steering towards Portoferraio'.

It was an astonishing mistake for Adye to make, especially when the visibility was very good and he had ample time to make up his mind. The log of the *Partridge* notes that as early as eleven that morning his look-out had seen 'three sail'. It also shows that before the breeze backed to the north-east, and he turned for Elba, Adye had come within a dozen miles of the convoy. What were the 'three sail'? And why did Adye fail to identify the *Inconstant* when Napoleon's men had so easily picked up the *Partridge*?

The 'three sail' may have been some English and Swedish vessels which had left Livorno on Sunday morning, or Adye may have taken the *Inconstant* and two of her consorts for merchant ships of this kind. Both are plausible suggestions. But there is a third and intriguing possibility. The three ships could have been the *Inconstant*, the *Melpomène*, and the *Fleur de Lys*, separating after a secret rendezvous north of Elba on Sunday night. That is what Ricci believed. On 2 March, when he was interrogated in Florence by Lord Burghersh, he stated that 'the French ships were with Bonaparte's fleet on Monday, and all were sailing together'; and Burghersh reported to London that when Ricci was cautioned 'against deceiving himself and the world upon a matter of so much consequence' he insisted that he had clearly seen the ships from above Portoferraio. Nicolson Stewart, an English tourist on Elba who watched the escape, also suggested that there had been some collusion with the frigates. When he arrived in Livorno on 3 March he declared that 'a French vessel of a smaller size came close into Elba on Saturday last, and made a certain signal which determined Napoleon to embark'. And within a few days Campbell himself came to the conclusion that either the frigates were curiously ignorant of the escape 'or else they were accessory to it and pretended ignorance'.

It may be as foolish to charge the French captains with complicity in the escape as to suggest that Campbell encouraged it, or that Adye deliberately turned away when he saw the *Inconstant*. But the mystery is compounded by the absence of the frigate logs, by the subsequent failure of the captains to explain exactly where they were or what they were doing during that critical weekend, and by Napoleon's own comment on the sympathy he could expect from the officers and men of the *Melpomène*. Perhaps the captains were unlucky or careless, perhaps they were simply concealing their failure to realize that the *Inconstant* was on more serious business than usual. Perhaps they had guessed what was happening and thought it too risky to ask their crews to seize the Emperor, or they had come to some covert understanding with him. No one knows. The fact is that the frigates which Beugnot had sent to prevent Napoleon returning to France proved to be no impediment at all.

Adye's failure to identify the *Inconstant*, at the moment when luck had put the Emperor at his mercy, was a more simple matter, which may be explained by a single sentence in Campbell's journal. 'In the course of the day', he wrote, 'we saw the French brig *Zéphir*.' It was almost certainly the *Inconstant* which they saw without realizing it. At a range of over twelve miles the two brigs, built in the same yard, would look very like each other. Adye and Campbell, moreover, were expecting to see the *Zéphir*, which Mariotti had just told them was on the way from Toulon to strengthen the watch on Elba, and they were half-convinced that if Napoleon had already sailed off in the *Inconstant* he would be heading for Naples rather than France.

Napoleon's luck and judgement had helped him to the narrowest of escapes.

★

Before Napoleon cleared the islands, however, there was to be another puzzling encounter with a French warship. The story of Captain Andrieux and the *Zéphir*, indeed, is even more curious than the failure of the frigates to intercept the convoy, for Andrieux—who was on his way to get 'particular instructions' from Mariotti about 'a cruise of observation' in Elban waters—actually came so close to the *Inconstant* that he could conduct a shouted conversation. According to Napoleon he was only 'half a pistol shot away'.

The remainder of Napoleon's account was thoroughly confusing. His statement that the meeting took place on Tuesday evening, when the flotilla was well past the tip of Corsica, was presumably a slip of memory, for Andrieux saw Mariotti in Livorno on Tuesday morning and reported that he had spoken to the *Inconstant* 'near Capraia' on the previous day. Napoleon may even have been telling something like the truth when he recalled that on sighting the *Zéphir* his naval officers wanted to board the brig and give its captain the chance of joining the expedition, and that he rejected the proposal because so little would be gained by success that it was not worth the risk of a failure which would jeopardize the whole venture. The rest of his account, however, wavered obscurely between a claim to have deceived Andrieux and an admission that the French captain had a very good idea what was going on and that he carefully avoided being involved in it. For Napoleon began by saying that Andrieux had innocently hailed the *Inconstant* and asked for news of Elba, that he failed to notice the grenadiers on deck because 'they had been ordered to take off their bearskin bonnets and tie scarves round their heads to look like sailors', and that Andrieux then resumed his course for Livorno—an improbable tale that others embroidered with a romantic description of Napoleon lurking behind the wheel-house and prompting Chautard's disarming replies. Since Andrieux had been close enough to speak and to hear he was bound to see that the *Inconstant* was heavily loaded and crowded with men, and in Napoleon's final paragraph, without any sense of contradiction, he quoted Andrieux's admission to that effect. Andrieux said later that the sight of so many men and small boats on deck aroused his suspicions; that these had been con-

firmed when he saw the other craft packed with troops; that he suspected that the Emperor was on board and bound for Italy; and
that if he had known that the flotilla was bound for France he
would immediately have hoisted the tricolour flag and joined it.

The most revealing remark in Napoleon's account, which squares
with what he had previously said about the crew of the *Melpomène*,
was the statement that Andrieux was 'one of the officers in the
French navy on whom the Emperor could most rely'; and Andrieux
was promoted after Napoleon's return to Paris. Whatever the
reason for Andrieux's failure to react it was certainly an encouraging augury of what might happen once Napoleon landed in France.
Mariotti, like other royalists, might huff and puff. 'If the *Zéphir*
had been here forty-eight hours sooner', he wrote to Talleyrand's
assistant Dalberg in Vienna, 'I should have given such instructions
that the *Inconstant* would either have been captured or sent to the
bottom with her cargo.' When it came to the point, however,
Napoleon showed a ghost-like ability to pass through any line of
watch which the Bourbons sought to place in his path.

★

Campbell and Adye were still out of luck, for the *Partridge* had lost
the wind entirely, and at eight on Tuesday morning she was lying
in a flat calm about four miles north of Portoferraio. From that
position they could not even see whether the *Inconstant* was there,
because the high forts were between them and the anchorage. But
they were now so convinced that something was wrong that Adye
thought it wise to keep the *Partridge* away from the fortress guns
while Campbell was rowed ashore to see what was happening. If he
failed to return to the boat within two hours Adye was to assume
that he had been arrested and to get a message to Burghersh in
Florence as quickly as possible.

It was ten o'clock when Campbell entered the harbour and he
saw at once that the sentries were men from the National Guard,
that all Napoleon's vessels had gone, and that the town was
strangely empty. Soon after he landed he met Henry Grattan, a
British tourist who had seen the flotilla sail on Saturday night,
though he had no idea where it was going. 'Some spoke of Naples
and Milan,' he told Campbell, 'others of Antibes and France.'
Campbell then went to see Fanny Bertrand, who said she knew nothing, but was in such a worried state that Campbell tried to shock
her into an indiscretion by saying that the plot was known and that

the fugitives were bound to be taken by the British squadron which was 'looking out for them' between Elba and Naples. 'On this she became more relieved and quite collected', Campbell remarked, 'from which I concluded that her opinion of their destination was north, and not south, as I thought at first.' At that moment, in fact, Napoleon was already south-west of Savona on the Ligurian coast.

Campbell next tried to bluff Dr Lapi, who boldly announced that he was governing the island on behalf of the Emperor, by asking for its surrender 'to the British, the Grand Duke of Tuscany, or the Allied Sovereigns', and when Lapi coolly ignored the demand Campbell brusquely announced 'that the island would now be considered in a state of blockade'. It was a rash threat for him to make on his own responsibility, for it was equivalent to a declaration of war long before anyone knew where Napoleon had gone or what he intended to do. It can only be explained by the fact that Campbell was beside himself with rage, or by some undisclosed knowledge that the Allies would treat any escape as the signal for a new conflict.

As Campbell passed the Vantini house on his way to the quay he called to ask if there was any service he could offer Napoleon's mother and sister. To his surprise he was invited into an antechamber, where Pauline seized the chance of an oddly flirtatious interview to cause delay and confusion. 'She then came out and made me sit down beside her,' Campbell wrote, 'drawing her chair gradually still closer, as if she waited for me to make some *private* communication.' She talked about her husband, and of her desire to go to Rome. She declared that she had had no knowledge of Napoleon's plans; and, Campbell added, she 'laid hold of my hand and pressed it to her heart, that I might feel how much she was agitated' by his departure. He decided that he was being teased and that he might usefully try another of his simple psychological tricks to discover where Napoleon had gone. 'During this conversation she dropped a hint of her belief in his destination being for France,' he said, 'upon which I smiled and said, "O non! ce n'est pas si loin, c'est à Naples", for I fancied (for the moment) she mentioned France purposely to deceive me.'

Campbell had at least learnt enough to change his mind and make him eager to get away. Collecting Grattan and Ricci, and requisitioning two fishing-boats to carry them over to the mainland, he went back to the *Partridge* to hold a council of war and to write his urgent despatches. Grattan was to go to Livorno, taking reports for Castlereagh, the British army and navy commanders in Genoa, and the ministers in Vienna and Paris. Ricci was to cross to Piombino and take duplicates to Burghersh in Florence. And the *Partridge* was to sail for Antibes.

That had been a difficult decision, and the four men had taken a long time discussing it. Grattan and Ricci seem to have shared the general opinion in Portoferraio that the Emperor had left for France: on the previous day the Oil Merchant had heard people in the town saying 'that the expedition was headed for Fréjus'. Adye, however, remained certain that Bonaparte had gone to join Murat, and in the end Campbell had to insist that the *Partridge* must sail north. 'There was always a probability of overtaking Napoleon and his flotilla, if he had gone in that direction,' Campbell wrote; 'there was none if he had gone to Naples.' There was other circumstantial evidence as well. If Napoleon had gone to join forces with his brother-in-law he would not have needed to take horses and cannon, or so much by way of stores and munitions, or so many civil followers and all the Corsicans. And there was Grattan's statement that he had watched the little fleet sailing northwards until late on Monday afternoon. With 'every minute of the utmost consequence', Campbell concluded, Napoleon would not have spent at least a night and a day working slowly to the north if he proposed to turn on his tracks.

'I think his destination is for the frontier of Piedmont next France', Campbell noted, 'and that he will take possession of some strong place near Nice, or between that and Turin, dispersing his civil followers immediately over North Italy, of which he will proclaim the independence, raising the disaffected there, while Murat does the same in the south.' Even when Campbell had rightly concluded that Napoleon was heading north the obsession with Italy exerted such a strong pull on his mind that he did not seriously weigh the attractions of a return to France against a harum-scarum intervention in Italy. Despite his long stay in Elba he so miscalculated the mood of Napoleon's men that he thought they would consider a landing in Italy 'less hazardous than raising the standard of rebellion in France, where they would be considered traitors'.

It was past two o'clock on Tuesday when the *Partridge* cleared Portoferraio and headed for France. Although Napoleon had a start of forty hours the winds were so uncertain that there was a chance of overtaking his convoy, and within twelve hours the sloop was fifteen miles north of Cap Corse and making a good eight knots an hour. Suddenly a light was sighted. 'We beat to quarters,' Campbell noted, 'as it was reported that there were several sail.' He was disappointed. The *Partridge* had met the *Fleur de Lys*.

Adye and Campbell went across to talk to Captain de Garat. 'He did not know of Napoleon's escape till we informed him,' Campbell wrote disparagingly, 'although his only duty was to prevent it, and he ought to have been off Elba as a watch, unless he was accessory to it.' No doubt de Garat told Campbell much the same story as he told to an official enquiry in August. Part of it was a blustering defence of his loyalty to the Bourbons. Part of it was an excuse. The *Fleur de Lys*, de Garat said, had been at sea in bad weather for six weeks, and in these conditions nothing but 'a stroke of luck' could have enabled the frigate 'to stop or even to sight Bonaparte's brig'. And part of it was a forceful and misleading claim that the *Inconstant* and the rest of the flotilla could not have passed Capraia, because he was lying to the north-west of the island all Monday afternoon and north-east of it on Monday night and Tuesday morning. Since Napoleon undoubtedly sailed north of Capraia on Monday afternoon de Garat was not telling the truth. Either he had been away from his station and had let the convoy through by default, or he had been where he claimed to be and let it pass by design. But whatever his reason for deceiving Campbell his story had serious consequences.

From five on Wednesday morning until seven that night the *Fleur de Lys* and the *Partridge* sailed on together towards France, while Campbell tried 'to reconcile Chevalier de Garat's information with that of Mr Grattan' and to decide where to look next for Napoleon. Small events may change the course of history. It was this encounter with the *Fleur de Lys*, Campbell wrote a week later, 'which alone prevented the *Partridge* from arriving at Antibes nearly about the same time with Napoleon, and lost us the glorious chance, which was so nearly at our command, of destroying him'. Campbell now decided that Napoleon 'may have secreted himself for a few days in Capraia or Gorgona, in order to lead away the *Partridge*, and be able at night to take Livorno by surprise'. He therefore turned the *Partridge* back to search the islands again, leaving the *Fleur de Lys* to go on to France with despatches for Paris and for Castlereagh. The fear of 'a ready communication with Murat' thus served Napoleon as a diversion to the very last moment, for it was this which made Campbell abandon the chase when there was still time for the *Partridge* to catch the *Inconstant* and the flotilla, or at least to get to Antibes before the column march- ed inland. As the sloop slowly worked its way back towards Tus- cany he was afraid he would find that Napoleon had already joined up with Murat's army on the way to Florence or that Murat's fleet lay between Portoferraio and Piombino.

The facts were very different. Soon after the *Partridge* met the *Fleur de Lys* the Emperor and his convoy were off Monaco, less than ninety miles away, and only a few hours from Antibes.

★

On Tuesday afternoon, as the sun caught the snow-capped peaks of the Maritime Alps, each man could see for himself where the flotilla was heading. 'The Seine water will soon cure you', Napoleon said jovially to the seasick Peyrusse, promising that they would both be in Paris by the King of Rome's birthday on 20 March. He gave the Legion of Honour to Captain Chautard and Lieutenant Taillade as a reward for the safe passage from Elba, and a strip of red bunting was cut up to make the same decoration for the hand- ful of *grognards* who still lacked it after years of campaigning. He talked optimistically to his officers. 'There is no precedent in his- tory for what I am about to do', he told Colonel Mallet of the Guard, 'but I can count on popular astonishment, the state of pub- lic opinion, the resentment against the Allies, the affection of my soldiers, and the attachment to the Empire which lingers every- where in France.' He teased the gloomy Drouot. 'I know that if I had listened to our wiseacre I should never have started', he said, 'but there were even greater dangers at Portoferraio.' And when a few members of his personal staff seemed to share Drouot's uneasi- ness at this 'bold and unexpected stroke' he dismissed their fears. 'I shall arrive and find no organized resistance,' he insisted, repeating the claim he had made to Drouot when he first told him of the plan to escape; 'I shall reach Paris without firing a shot.'

The same note of lofty confidence ran through the proclamations which Napoleon had printed before he left Elba. The first was his appeal to the people of France. It went over familiar ground, claim- ing that Napoleon was still undefeated when he was betrayed by Augereau and Marmont a year before, that the Allies had seized their chance to impose a humiliating peace, that the Bourbons— who had learned nothing and forgotten nothing—were trying to restore the privileges of the old regime, and that Napoleon was returning to promote peace in Europe and a happier future for his countrymen. But for all its familiarity it caught the mood of the time exactly. 'Frenchmen,' Napoleon declared in a characteristic blend of imperial bombast and republican rhetoric, 'your complaints and your desires have reached me in my exile. You have asked for the government of your choice, which is the only legitimate government. I have crossed the sea, and I am here to resume my rights, which are also your own.' Constitutional niceties had never been Napoleon's strong point when he was reaching for a throne.

The second manifesto, addressed to the veterans of the Grand Army, made the same point in more stirring language to the men who had followed Napoleon from one battlefield to another over the past twenty years. 'Soldiers,' Napoleon exclaimed. 'Rally round the standard of your chief. He lives only for you, his rights are only those of the people and your own. Our victory advances like a charging line of battle! The eagle shall carry the tricolour from steeple to steeple until it reaches the spires of Notre-Dame!'

The third document was supposed to be an appeal from the men of the Guard who had accompanied the Emperor into exile, and to give it an appearance of spontaneity Napoleon produced a manuscript draft, rather than the printed version, and he set every man who could write to make copies of it from dictation. 'Friends and comrades-in-arms, return to your duty,' it said. 'Trample the white cockade, the badge of shame!' At dawn on Wednesday morning, with France in sight, Napoleon ran up the tricolour flag, and the *grognards* followed his example, taking their treasured bunches of red, white, and blue ribbon from their packs and proudly pinning the faded emblems to their bonnets again.

It was one o'clock on the first afternoon in March when the flotilla passed Antibes, turned in beyond the short peninsula and anchored at Golfe Juan, about a mile to the west on the road to Cannes. Napoleon immediately sent Captain Lamouret ashore with twenty men to neutralize the Gabelle fort which dominated the beach, and to tell anyone they met that the troops were simply a large party of the Guard returning from Elba on leave. The first precaution was unnecessary, as the fort was unoccupied. The second was temporarily convincing, since several parties of men on leave had landed at Antibes in recent weeks, and that confusing report ran ahead of the news that this time the Emperor had come with them.

Although Lamouret was quickly followed by Cambronne, with a hundred picked grenadiers who were to cover the beach and then to form an advance guard for the march, the landing itself was a slow and makeshift business, and the band amused the toiling men by repeatedly playing the popular and appropriate tune 'Where Can One Better Be Than in the Bosom of One's Family?' It was dawn on Thursday morning before all the troops, stores, the treasure, the cannon, the horses, and the carriage were safely unloaded. It would have been a most dangerous business if the *Partridge* or the *Fleur de Lys* had come up in time, or there had been any opposition ashore. Two reliable companies of the 106th Regiment, which was exercising outside Antibes that afternoon, would have been quite sufficient to stop the boats reaching the beach. Yet nothing was done. When the zealous Lamouret took his men on to Antibes to demand the surrender or defection of the garrison the major who was in command played safe. Perhaps he thought that his men were not reliable. He was content to arrest Lamouret's little squad, to send a courier to General Corsin who was away on a visit to the nearby island of Sainte Marguerite, and to shut the town gates. At the same time Cambronne cut the road leading west from Antibes and moved on to Cannes without encountering any resistance.

Soon after four o'clock Napoleon himself was rowed in from the *Inconstant*. He walked up the beach to the olive grove where his troops were bivouacked for the night, and sat by a camp fire chatting to soldiers and some peasants who had come to see what was happening. The eagle had landed again in France, and with the dawn he would begin to fly northwards.

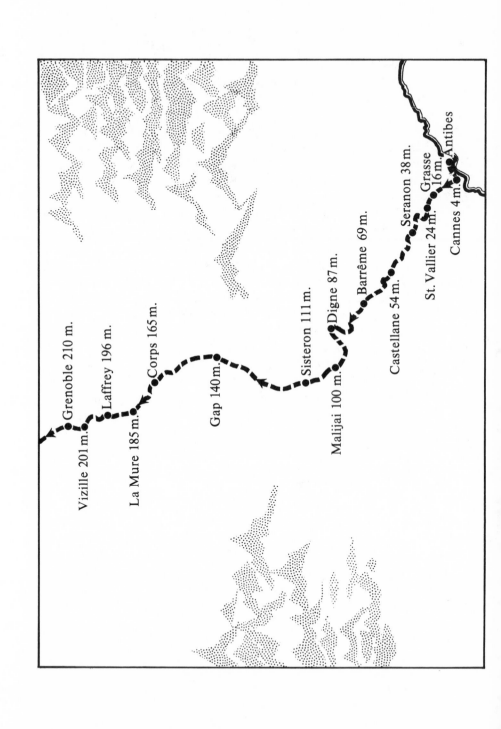

Grenoble 210 m.
Vizille 201 m.
Laffrey 196 m.
La Mure 185 m.
Corps 165 m.
Gap 140 m.
Malijai 100 m.
Sisteron 111 m.
Digne 87 m.
Barrême 69 m.
Castellane 54 m.
Seranon 38 m.
St. Vallier 24 m.
Grasse 16 m.
Cannes 4 m.
Antibes

19

THE TEST OF SUCCESS

The failure to secure Antibes was a disappointment, but Napoleon refused to attack the town where he had been briefly imprisoned after Robespierre's fall in 1794, or even to waste a few hours seeking the release of Captain Lamouret and his men. 'Time is too precious,' he said when some of his officers proposed to make a fight of it. 'The only way to correct the bad impression created by this affair is to travel more quickly than the news. Why, if half my force were prisoners at Antibes I should leave them there. If they were all there I should go on alone.' This decision was not merely a matter of time, although Napoleon knew that he would have to march his men more than thirty miles a day if they were to cover the six hundred miles to Paris by his son's birthday on 20 March, and that even his seasoned *grognards* could not move at such a speed if they ran into any resistance on the way. It was really a matter of personal magic. Napoleon believed that his return would cast a spell on France and that the first shot would shatter it.

Cannes was the first test. As soon as the moon was up Napoleon led his men along the coast to join Cambronne's advance guard, which had moved into the village in the late afternoon without meeting trouble or enthusiasm. The inhabitants, indeed, seemed to be stunned by the sudden appearance of light infantry sporting the tricolour cockade; and in the small hours of the morning, when Napoleon arrived, they were still standing by the bivouac fires in front of Notre Dame de Bon Voyages in a state of anxious curiosity. It was a dispiriting moment, and Napoleon had to get what comfort he could from a friendly chat about conditions in Paris with his former equerry, the Prince of Monaco, who had been detained by Cambronne as he passed through Cannes on the way home. Then, to everyone's surprise, Napoleon left the main highway that led to Aix and led his troops up through Le Cannet on the rough carriage road which climbed fifteen miles through the hills to Grasse.

Once again there was no opposition. As soon as the royalist mayor of Grasse heard of the landing he vowed to defend the town, and he asked the retired General Gazan to help him. But the general was a realist. Discovering that there were only five usable muskets in the place, and that no one was volunteering to carry them, he told the mayor to cool his martial ardour, sent a message to the War Ministry in Paris, and rode off in the hope that he would meet the soldiers from Antibes in hot pursuit of the Emperor. It was a vain hope, of course, because nothing had been done to despatch the garrison from Antibes. Nothing effective was to be done anywhere else in the next few vital days, partly because the authorities were both surprised and ill-prepared, partly because they feared the line regiments were too disaffected to be dependable, and partly because there were so many men in official posts who decided that delay was the better part of valour when they learned that Napoleon was back in France. Fearing to compromise themselves by declaring firmly for Bonaparte or the Bourbons, and wishing that someone else might settle the embarrassing problem for them, these men who had made their career in the Empire danced a hesitation waltz while the Emperor marched.

The situation in Grasse was typically ambivalent. The town would neither resist Napoleon by force nor welcome him with cheers. Early on Thursday morning, when Cambronne rode in to demand rations for a thousand men, the mayor preferred rhetoric to heroism and 'many old heads and white ribbons' filled the square while he bandied bold words with Cambronne. Other townspeople streamed out to stare at the Emperor and his men, who had halted above the town on the nearby plateau of Roquevignon. There was wine for the guardsmen, a few cries of 'Vive l'Empereur!' and some people, it was said afterwards, brought symbolic bunches of violets. But no one seemed keen to follow the eagles on the march to Paris. The only recruits on the first day were two deserters from Antibes and a local tanner.

There was bad news, too, at Grasse, for Napoleon now learnt that the road on to Digne and Sisteron which he had authorized in 1809 had never been built. Unless he went back to the main route through Aix-en-Provence, losing time and also moving dangerously close to the large garrisons at Toulon and Marseilles, he would have to abandon his cannon, Pauline's carriage, and sixteen waggons containing supplies. The one possible route to the north was by a narrow snow-bound track through the mountains, where in some places he and his men could only walk in single file and it would be necessary to pack the treasure and other baggage on a string of mules. 'Our little force', Major Laborde wrote afterwards, 'stretched as far as a column of twenty thousand men on a normal road.'

Napoleon allowed his men to rest at Grasse from eight o'clock until noon, and then he set off in the cold bleak landscape of Haute Provence. He crossed the Col de Pilon at 2,500 feet, dropped down into St Vallier de Thiery, and climbed up over the Pas de la Faye. Long after darkness fell a straggling line of soldiers followed him into the hamlet of Seranon, set in a broad valley bordered by high rock walls, 3,000 feet above sea-level and thirty miles from Cannes. It had been the worst part of the journey. In the long ravines near Escragnolles the soldiers had been forced to scramble as best they could up steep defiles and along ravines rimmed with ice. Napoleon kept up with them, plodding on rather breathless, for no one could ride. Even a sure-footed mule, carrying part of the treasure, had slipped to its death in a deep gully, and two thousand gold napoleons were lost in the rushing water.

Despite these arduous conditions, however, Napoleon was determined to keep to his timetable, and he led his men off to an early start on Friday. By the middle of the day they reached Castellane. This little town at the western end of the Verdon gorges was so isolated that the appearance of Cambronne, with a demand for five thousand rations of meat, bread, and wine, was the first the townspeople knew of the Emperor's approach. As they scurried to find these supplies, and fresh mules to carry them, Napoleon received the mayor and secured some blank civilian passports. One was for Apollinaire Emery, the surgeon from Grenoble who had secretly travelled to Elba in September with young Dumoulin. The fate of the whole expedition was likely to turn on the speedy seizure of Grenoble, and Emery was being sent ahead to tell the Bonapartists of his home town what Napoleon expected of them. Pons was also given a passport and a mission. He was to turn off at Digne to take a message to Marshal Massena in Marseilles, urging him to declare for the Emperor.

The column made another thirty miles over the high pastures to Barrême that day. There was another early start on Saturday, and when Napoleon came down into Digne about ten that morning he found that this time he was expected. Two days earlier, in fact, the Prefect of the Var had sent a courier to tell General Loverdo, who commanded the small garrison of the Basse-Alpes, to hold the important bridge over the Bléone, and to prevent Napoleon moving down the road which led one way to Aix and the other to Sisteron. Once again discretion prevailed over duty. Loverdo did not have enough troops to hold the bridge for long, and those he had were of such doubtful loyalty that he had not dared to carry out some recent court-martial sentences on soldiers who had shouted Bonapartist slogans. And the leading citizens, afraid of reprisals if he tried to make a stand, persuaded him to take his troops away into the hills until Napoleon had come and gone. Then, with a clique of Bonapartists to encourage them, the townspeople gave the Emperor a cordial though cautious reception. A squad of mounted gendarmes joined him; so did a few veterans of the Grand Army; and while Napoleon rested and dined at the inn called Le Petit Paris a printer ran off hundreds of copies of his proclamations.

Ninety miles from the coast the column was at last marching on a passable road, and it made good time to Malijaï. While Napoleon halted at the château for the night, Cambronne and forty men hurried on for another twelve miles to seize the crossing at Sisteron, where the Durance ran between high cliffs to form a natural gateway out of Provence.

At one o'clock in the morning they arrived to find that the bridge was intact and that the fort which dominated it was empty. General Loverdo had previously thought of hurrying on to destroy the bridge, but he lacked the power to blow it up himself and his experience at Digne had shown that the local people were very reluctant to damage anything for the sake of delaying Napoleon for a few hours. In any case it would have been a useless gesture. Napoleon's marines could easily have made rafts from the timber which lined the river and ferried everyone across before any regular troops arrived to prevent them.

For Napoleon had stolen two marches on his pursuers. Marseilles was only eighty-eight miles to the south of Sisteron, ten miles less than the distance from Cannes, and the road was easy; but Napoleon was already at Barrême when Massena heard of the landing, and though he immediately sent General Miollis with two regiments to block the road to Grenoble he had served Napoleon long enough to suspect that this was a lost cause. Unless the Emperor had run into difficulties in the mountains he was bound to be miles ahead of Miollis at Sisteron, and still gaining on him.

Napoleon was also gaining ground politically. After Digne he had put his army into proper marching order for the first time. The leading detachment consisted of three companies of light infantry, with the marines and the Polish lancers, who had been acquiring suitable horses along the route. Then came three companies of grenadiers, Napoleon, his staff, the treasure, and the three hundred Corsicans who formed the rearguard. As a fighting force, if it encountered any serious opposition, this column had little more value than a large escort of gendarmes, but its emotional appeal was out of all proportion to its size. There was a lukewarm welcome at Sisteron where Napoleon lunched at the Bras d'Or, but on Sunday afternoon, as he pushed on past the village of Poët into the traditionally republican districts of the Dauphiné, the peasants were at last coming to stand by the side of the track and cheer with the old fervour. They fetched out veterans to see the soldiers pass, and to confirm that the little man in the grey riding coat was indeed 'Corporal Violet', who had been expected back with the spring flowers; and where there was no one who knew him by sight they took five-franc pieces from their savings to compare the image on the coin with his profile.

At Gap, set in a great bowl of snow-covered mountains, the *grog-nards* who had marched 150 miles joined the townspeople in the square, singing the 'Marseillaise', the 'Ça Ira', and the 'Carmagnole' as if the revolution were beginning all over again. As Napoleon slept in the Hotel Marchand it seemed that he had gone to Elba as a defeated Emperor and had come back as a triumphant Jacobin, ready to drive the Bourbons out for the second time.

*

While Napoleon led his soldiers quickly over paths more suitable for shepherds and smugglers the unfortunate Campbell was wasting his time on a fool's errand. The wind was so slack that he did not get back to Capraia until early on Friday morning, and as soon as he landed he was told by the commandant and the mayor that Captain de Garat had misled him. They had definitely seen Napoleon's flotilla pass the island on the Sunday afternoon. 'Captain Adye, therefore,' Campbell wrote in a tone of disconsolate weariness, 'has again shaped his course for Antibes.' By then the search was degenerating into complete futility. On Saturday evening the *Partridge* met the *Wizard* which was making for Livorno, and Campbell sent a message to Burghersh confessing that he still had no idea where Napoleon had gone. On Sunday that puzzle was solved, for an Italian vessel coming out of Genoa reported that Napoleon had landed at Antibes, but Campbell continued to fret as the *Partridge* lay idle in the good weather, and the crew were reduced to 'drawing and knotting yarns' and setting variations of sail in the hope of catching the light airs. It was clearly a trying time for everyone on board. 'Punished Alex. Stevens with 36 lashes for striking his superior officer and disobeying orders,' Adye wrote in the ship's log, 'also George Crawley and

William Poole with 12 lashes each for disobedience of orders.'

The *Partridge* drifted slowly down the coast, with each day bringing an instalment of bad news. On Monday a British troop transport confirmed that Napoleon had 'marched into the interior'. On Tuesday the sloop met the *Aboukir* and its captain told Campbell that the Emperor was well on his way to Grenoble 'without opposition'. When the *Partridge* finally reached Antibes, a full week after the landing, Campbell found the *Fleur de Lys* lying idle 'in the offing', and General Corsin busy explaining why he had been away from his command until after Napoleon had moved on. Tactfully, Campbell made no attempt to charge de Garat with misleading him, and he allowed himself only one pointed comment on General Corsin and the Antibes garrison. 'It seems most extraordinary', he wrote, 'that the disembarkation should have taken place, and the encampment continued from midday until almost sunrise next morning, without attracting more notice, or causing any measures to be taken on the part of the authorities.'

Uncertain what to do next, and eager to pick up snippets of the news that was now filtering back to the coast, Campbell hung about for a couple of days. None of this news was good. The French consul in Nice regretted 'that no opposition had been made in Napoleon's front during his march'. The Governor of Nice complained 'that the spirit of France is not so good as might be expected . . . the people are stupefied'. The men from the Guard who had been taken prisoner at Antibes 'were treated like friends, and were seen playing bowls with the garrison'. And the speculation was even more depressing. 'It is probable that the test of Napoleon's success will be made at Grenoble', Campbell noted on 10 March, 'and that he will endeavour to bring it to that issue as soon as possible before the accumulation of force renders his passage of the Isère more difficult.' Anxious to obtain what he called '*circumstantial facts*' for himself, Campbell belatedly set off into the mountains of Provence, picking up more worrying anecdotes about Napoleon's march and the failure of the authorities to obstruct it.

At Draguignan, where Campbell was staying with the Prefect of the Var, he was roused at one o'clock on the morning of 13 March with the news he had feared. Napoleon had entered Grenoble six days before, and the troops sent to stop him had 'failed in their duty'. Campbell turned round again, this time to seek Lord William Bentinck in Genoa to tell him of 'the little reliance that can be placed on the regular army, so that he may prepare for the worst'.

He did not know what else to do, and he now realized that the mission which had begun at Fontainebleau almost a year before was ending in confusion and possible disgrace. When he returned to Nice he found that Lord Sunderland had just come in from Marseilles with a most disconcerting report. 'There it is universally believed that the English had favoured Napoleon's return, and the people are furious against us,' Campbell noted. 'The same idea also prevails everywhere in the south of France and in Piedmont. A newspaper in Turin, just arrived at Nice, states positively this to be the case!'

On 10 March, when Campbell wrote to Lord William Bentinck from Nice, he confidently declared that, under Castlereagh's authority, he proposed to 'act in regard to Napoleon until his career is terminated, which I trust will be soon'. But when he heard that Napoleon was already at Grenoble his morale seems to have collapsed. The change showed immediately in his diary, which was suddenly reduced to a perfunctory record of his return to England. Even that journey had its troubles. He left Genoa on 20 March. 'During the night robbed of my watch and between fifty and sixty guineas by bandits near Novi.' On 23 March, at Vevey: 'Carriage wheel broke'. The laconic entries went on for the next ten days as Campbell travelled down the Rhine in long stages and took the packet from Ostend to Deal. On 1 April he arrived in London. Next day he was received by Castlereagh, and then by the Prince Regent who was 'pleased to express in unqualified terms his entire approval of Sir Neil's conduct . . . in no way would he, as English Commissioner, be considered responsible for the unfortunate evasion of Napoleon'. The government was to stand by its man on Elba, as its critics in Parliament and the newspapers tried to find a scapegoat, and as rumours of Campbell's complicity flew about Europe as if they were proven fact. Yet Campbell knew that he would never live down 'the violent and general' prejudice against him. It was his misfortune that without suitable experience, clear instructions, and adequate naval support, he had been sent to keep watch on a man with the luck of the devil. The astonishing thing is not that he failed, but that for all his disadvantages he came so close to succeeding.

★

There had been so many rumours that Napoleon was about to escape from Elba that no one in Vienna took them very seriously.

The police inspector who was spying on Marie-Louise at Schön-brunn, for instance, noted that the mood of the remaining French members of her retinue had changed. 'They are saying that it won't be long before the bird takes flight', he reported to Baron Hager on 13 February, 'and that the English will be unable to do anything about it.' Another informer declared that 'the hour has struck for Murat', and a third said that Murat would soon call on 'all the hot-heads in Milan, the rascals, the young enthusiasts, the ideologues, Napoleonists, etc. to make him "the King of all Italy"'. The fail-ure to take alarm was foolish, yet understandable. For one thing, rumour was more often wrong than right; and for another, the Congress was now moving quickly towards a general settlement, and the other sovereigns seemed increasingly disposed to let the Austrians settle the Italian question when everyone else had gone home.

Their dismayed surprise was thus all the greater when Metter-nich delivered the news he had received from Campbell early in the morning of 7 March. 'I dressed myself in a flash,' Metternich recall-ed, 'and before 8.0 a.m. I was with my Emperor.' Francis reacted calmly. 'Napoleon appears anxious to run great risks; that is his business,' he declared. 'Our business is to give the world that re-pose which he has troubled all these years. Go at once and find the Emperor of Russia and the King of Prussia; tell them that I am prepared to order my armies once again to take the road to France.' Before ten in the morning Metternich had seen both monarchs, dis-cussed the situation with Field Marshal Schwarzenberg, sent mes-sengers to halt the withdrawal of the Allied armies, and called Wel-lington, Talleyrand, Hardenberg, and Nesselrode to an emergency meeting. 'In this way', he noted 'the decision to make war was taken in less than an hour.'

The effect on the leaders of the Congress was so startling, in-deed, that the newspapers were forbidden to report the escape for several days while the sovereigns and their ministers considered what they should do. 'Happily it has happened while we are still here at Vienna', Francis said in a tone of relief. If the Congress had already completed its work, if the sovereigns had gone home, and their armies stood down, Napoleon could more easily have set the former allies at odds with each other and greatly enhanced his chances of survival. As it was the despatches that came in from Genoa provoked a round of recriminations. 'That's what happens as a result of all those useless and endless discussions at the Congress',

Talleyrand said waspishly. Like Castlereagh he thought that the trouble stemmed from the Tsar's hostility to the Bourbons and his indulgent attitude to Napoleon. And Castlereagh in turn was blamed for the Emperor's flight. 'There are those who believe that the English have allowed him to escape in order to recapture him or to have an excuse for treating him with the utmost severity,' an informer told Baron Hager on 8 March; 'others say that it was not an escape at all but a departure arranged by the English who perhaps wish to send him to America.' The uncertain mood of the moment was well expressed by the King of Prussia. 'For a long time Napoleon has deserved to be hanged,' the French royalist Count Alexis de Noailles remarked to the King. 'Now that he has escaped it is essential that this time he really is hanged.' 'Perfectly true, my dear count,' Frederick William replied, 'but first you have to catch him!'

Once it was known that Napoleon was definitely in France the difficulty of catching him became apparent. 'He will head straight for Paris', Metternich said bluntly, when Talleyrand suggested that the Emperor might seek asylum in Switzerland, and the realistic voice of Metternich echoed in the letter about Napoleon's intentions which Friedrich von Gentz wrote to Count Munster of Hanover on 16 March. 'The situation is very serious,' Gentz insisted. 'The army is entirely for him. One cannot count on the National Guard. If Bonaparte succeeds, without any doubt he will immediately march on Belgium and the Rhineland.' A few days later Count Humbert de la Tour du Pin, a young French officer who was visiting Vienna on leave, asked his father why he was so surprised at Napoleon's easy progress: 'Haven't you been reading my letters during the past three months, in which I've been telling you about the discontent in the army?'

The prospect that Napoleon might soon be back in the Tuileries was a matter of deep concern for all the powers who had recently combined to defeat him, but it was a particular worry to the Austrians—partly because it would reopen the embarrassing question of his wife and child, and partly because they were on the point of agreement with Louis about Lombardy and Naples. Francis might be able to solve the domestic problem, because Marie-Louise was so deeply involved with Neipperg and anxious to be settled with him in Parma that she would resist any other proposal. Within a week of hearing the news from Elba she wrote to Metternich 'to say that she has never and will never take the least part in the plans, projects and enterprises of Napoleon, and that she agrees entirely with him about her interests and those of her son'. But the secret negotiations about the fate of Murat were a much more serious business. If the Bourbons were overthrown the plan which Metternich had concocted to consolidate Austrian rule in Italy would be completely ruined.

★

Castlereagh himself had taken that plan to Paris on his way to England at the end of February. He had been recalled because Lord Liverpool felt that he was needed to strengthen the government against its critics in the House of Commons, and that the Duke of Wellington could cope just as well with the remaining formalities in Vienna. Although Castlereagh disagreed, and argued that he had more important work to do in Vienna than in Westminster, he realized that by calling at Paris with the draft agreement he could ensure secrecy, demonstrate British approval for it, and see Louis to discuss the last serious impediment to signature, the fate of the Parma duchies, which Louis wanted for the Bourbon claimant and Francis needed for Marie-Louise. When Castlereagh talked to the King on 27 February they reached a compromise. Marie-Louise was to have Parma for life on condition that she renounced any other claims under the Treaty of Fontainebleau, and that it reverted to a Bourbon heir on her death instead of to the King of Rome.

The rest of the scheme was comprehensive and ingenious, for Metternich had seized his chance to turn the Bourbon obsession with Murat to Austria's advantage, and to get some compensation south of the Alps for the concessions he had been forced to make in Poland and Saxony. It bound France, Austria, and Britain to a common policy. It required the French to abandon their risky notion of a march on Naples which might precipitate a general rising in Italy and possibly lead to the disintegration of their unreliable army; to recognize that the peninsula was actually an Austrian sphere of influence; and to pay twenty-five million francs towards the cost of putting Ferdinand back on his throne in Naples. Since the British were pledged to provide the necessary naval support and supplies as the Austrian troops moved southwards, the Austrians were in effect undertaking to get rid of Murat on commission, and at their own convenience.

The price of the Bourbon attachment to family rights was high— at least ten times as much as the annual pension which France had failed to pay Napoleon on Elba. But Louis was so eager to pay it, and to see results for his money, that on 4 March, three days after Castlereagh had left for London, Blacas was sent 'to make the arrangements for the Naples business' with Field Marshal Vincent who was now acting confidentially for Metternich in Paris. If Vincent could offer some hope that Austria would move quickly, Blacas said, he could offer a down payment of ten million francs within two hours.

At three o'clock that afternoon the news of Napoleon's landing reached Paris. The one possibility that Metternich had not foreseen when he drafted the secret treaty was that the King of France would fall before the King of Naples. In the event Austria would have to fight Murat alone, and without Bourbon money.

★

The news of Napoleon's escape had an equally disturbing effect in Naples, where Colonna's arrival sent Murat into a state of scarcely controllable excitement, and set him quarrelling with his wife and his Council of Ministers. Convinced that Napoleon's gamble would end in disaster and that it could easily ruin Murat as well, Caroline persuaded the Council to declare that 'this event does not alter our policies', and she did all she could to support Napoleon's own proposal that Murat should remain calm and conciliatory towards the Austrians. 'The Emperor Francis has supported us loyally so far', she said, 'and I am sure he will not abandon us now *if we merit it.*' But Murat was not to be pacified. Like an old war-horse pricking its ears at the sound of a distant trumpet he was keen for action, and for glory—the seductive combination of vanity and power that drove Napoleon himself, and possessed almost everyone who served under him.

The Austrian envoy de Mier was well aware of this dangerous mood. 'Murat's uneasiness and bad humour increases from day to day', de Mier wrote to Metternich, 'and I fear that the hotheads who surround him will engage him in some false step.' And when General d'Ambrosio returned from the Congress and urged Murat to maintain his links with Vienna as a means of self-preservation, he was treated to a manic harangue. 'What need have I of alliances, since the Italians salute me as their sovereign?' Murat asked. 'Will not Austria soon be called upon to fight France. and how can she then face all ways and resist all her enemies at the same time?' When d'Ambrosio tried to reason with him Murat shouted back.

'We shall succeed. I shall call together all the people of Italy at Bologna. It's no longer time to negotiate. It is time to fight!'

On 9 March the *Inconstant* came into Naples with the news that the landing at Golfe Juan had gone off well, and that Napoleon had last been seen on the road to Grasse. This report roused Murat to a new pitch of rhetoric. 'At last everything goes marvellously well,' he wrote to Lucien Bonaparte in Rome. 'I have taken my side.' And he was even more ecstatic in a letter to Cardinal Fesch. 'I wish to prove to the whole world that I have never been, that I am not now the enemy of Napoleon.' Caroline tried again to control her husband. 'If he tries to fly with his own wings he is lost', she said despairingly to de Mier. But she could only delay the inevitable for a few days. Murat was bent on sharing Napoleon's triumph or dying in the attempt.

★

As Napoleon left Gap on the afternoon of Monday 6 March that triumph began to seem reassuringly possible. A cheerful crowd followed the column for some way out of the town, and all through the high wintry countryside, as Napoleon climbed up from one valley and down into the next, the peasants were waiting with weapons in their hands to offer help. At St Bonnet, for instance, the local people were so worried by the size of the little army that they wanted to ring the tocsin, rallying all the National Guard units in the neighbouring villages to strengthen it. But Napoleon tried to keep this tide of enthusiasm within bounds. He was counting less on the soldiers who had come with him from Elba, or on the peasants who seemed prepared to start another revolution, than on those he had yet to meet. Somewhere between Gap and Grenoble he was bound to face regular troops for the first time, and when that happened they would either be overwhelmed by the sight of the Emperor, or they would stand firm. If they stood their ground his adventure would come to an ignominious end.

At Gap, again, there had been no resistance. General Rostollant, the military commander of the Hautes-Alpes, had only a handful of men, some guns but no gunners, and he was as ambivalent in his loyalty as his officers. Although he made a show of activity by sending off forty-odd messages to other commanders in the region, Rostollant carefully kept out of Napoleon's way, and he defected within a week. On 10 April, in a neat explanation of the dilemma which faced so many officers on Napoleon's return, he said that he had 'done everything in the first embarrassing moments to reconcile his honour and his duty with the attachment and devotion that he had never ceased to feel towards the Emperor'.

Such luck, however, could not last. Late on Monday night Cambronne ran into a company of army engineers sent to destroy the important bridge at Ponthaut, where the river Bonne ran through a deep gorge. The engineers were in a disorderly mood; some of them had thrown away their white cockades and shouted 'Vive l'Empereur!' as they had marched up from Grenoble, and when they reached the small town of La Mure, about two miles short of the bridge, the mayor had easily persuaded them to abandon their task. It would be pointless to wreck the bridge, he said, in words which echoed the prevailing sentiment in Digne and Sisteron, since Napoleon could easily cross the river at a nearby ford.

Cambronne, pressing forward with the advance guard, climbed up to La Mure near midnight, and learnt that there were troops from Grenoble ahead. He thought of holding the town, which lay in a neck of the hills before the road to Grenoble began to run down through a broad valley and past a string of small lakes, but he was alarmed when he heard that there was an infantry battalion on the move nearby. Fearing to be outflanked he withdrew to Ponthaut, where he held the bridge and sent an urgent message to Napoleon, who had stopped for the night some miles back at the Hôtel de Palais in Corps.

The vital confrontation had come at last.

★

The odds seemed heavily against Napoleon. Beyond the troops who closed the road at La Mure lay the strong garrison at Grenoble, commanded by General Marchand, a local lawyer who had enlisted in the revolutionary army in 1791 and risen to field rank. Marchand had the 5th Regiment of infantry, the 3rd Regiment of engineers, and the 4th Regiment of artillery to hold the walled town, and he had already summoned the 7th Regiment from Chambéry and the 4th Hussars from Vienne. If the soldiers were only as grudgingly loyal to the Bourbons as those at Antibes, Napoleon might be bluffed into surrender, or withdrawal, and even one serious set-back might be enough to take the heart out of the rebellion.

But the men were not loyal. The garrison and many of the townspeople had resented Marchand's new-found royalist convictions; the 4th Artillery was proud to be the regiment in which Napoleon had started his military career, and the town was seething with Bonapartist intrigue. Dr Emery had already arrived with copies of the Emperor's proclamations, after being twice arrested on the way and twice released for fear of reprisals when the Emperor himself arrived; and he and his friend Dumoulin had started a clandestine campaign to corrupt the soldiers and rouse the civilians. By 5 March the signs of Bonapartist sentiment were so strong that General Marchand made his men repeat their oath of allegiance to the Bourbons, and listen to an order of the day in which he declared that Napoleon had no chance of success since he had only a thousand men. When this sullen ceremony was interrupted by cries of 'What about us, then?' and 'Don't we count, too?' Marchand decided that he must keep most of the garrison inside the walls to reduce the risk of desertion.

Next day Marchand heard that Napoleon was coming on fast. That was when he sent the party of engineers to blow up the bridge at Ponthaut, and a few hours later, almost as an after-thought, he despatched Colonel Lessard and a battalion of the 5th Regiment with rather ambiguous orders to support the demolition squad. It was never clear whether Lessard had been given authority to open fire if Napoleon continued to advance, and his will to resist was undoubtedly affected by the growing disaffection of his men. When they passed through the village of Vizille, where the meeting of the Dauphiné Assembly in 1788 had seen the beginning of the French Revolution, even the children were calling out 'Vive l'Empereur!' and the soldiers were muttering uneasily. 'They can do what they want', one man was heard to say, 'but we aren't going to fight our own mates.' It was a hard steep climb from Vizille, and Lessard only reached the outskirts of La Mure shortly before Cambronne. When he heard that he was in contact with Napoleon's advance guard he withdrew the engineers and the infantry through the village of Pierre-Châtel towards Laffrey, on the side of one of the lakes some miles back towards Grenoble. It was the sound of this movement which had worried Cambronne and made him fall back on Ponthaut.

Next morning, Tuesday 7 March, Cambronne returned to La Mure, found it empty of troops and full of cheering people, and left the Emperor to come up and enjoy the welcome while the advance guard pressed on to look for Lessard. Some five miles towards Grenoble the Polish lancers came across his battalion drawn up in battle order across the track, with its right flank against a fairly steep hillside and the left flank reaching to the lake. It was a strong position, and the forces were nearly equal, separated by something over a musket shot. They faced each other all morning. Lessard was waiting for a reply to the message he had sent to Marchand during the night, requesting further orders; Cambronne was waiting for Napoleon to finish his festive round in La Mure and come up to decide what should be done.

During this long delay Lessard arrested a lawyer who had ridden out from Grenoble and was trying to read Napoleon's proclamation to the men of the 5th Regiment, but he seems to have been unable to prevent peasants fraternizing with the soldiers or to do more than drive off some half-pay officers in mufti who were trying to discover whether the soldiers would really fire on the Emperor. The best he could do was to hold his ground, refusing to receive the orderly officers sent by Cambronne in the hope of enticing him to desert, and hoping that he would soon hear what Marchand expected of him. But when the royalist Captain Randon arrived from Grenoble about noon he could only give Lessard a vague order to resist. Marchand was either too irresolute or too anxious about the consequences to commit himself to anything more.

All the same the sight of Lessard's battalion was enough to worry Napoleon, who had come up in the course of the morning and seemed agitated by the difficult choice before him. He spent a lot of time looking at Lessard's position through his field-glass, conferring with Cambronne, Drouot, and Bertrand, and consulting with some of the men who had been moving between the opposing ranks. It was the critical moment of the whole march. There was another long delay, probably calculated, which made Lessard's infantry very restless, as if the presence of the Emperor and the Guard formed up behind the tricolour unnerved them. Lessard said afterwards that many of his men were ashy-faced and trembling, and in no condition to fight. And then, early in the afternoon, Napoleon sent his aide Captain Raoul forward with a final suggestion that Lessard should defect. When Lessard refused, Raoul shouted at his men, 'The Emperor is about to advance towards you. If you fire he will be the first to fall, and you will answer for it to France!'

The Polish lancers came on first, and Lessard ordered his men to fall back. Then, feeling that he had them under control, he turned and faced the cavalrymen again. At this the lancers swung away, revealing the long blue coats and bearskins of the Old Guard, with the Emperor among them, marching with their muskets under their left arms and their bayonets in their scabbards as a sign of peaceful intentions. As they came close they started shouting. 'We are Frenchmen too!' they cried. 'We are your brothers!' And then Napoleon himself called out to Lessard and his men.

There are several accounts of what he said. 'Soldiers of the Fifth,' the most likely version goes 'will you fire on your Emperor?' But it is doubtful whether anyone heard or remembered exactly, for Lessard's line was suddenly collapsing, and everyone was cheering, laughing, and crying at once. It was certainly an emotional moment. It was also rather less risky than legend and the devotional pictures suggest. Napoleon had plainly held back until the go-betweens assured him that Lessard's men were ready to break, and as they crowded round the Emperor many of the soldiers were at pains to tell him that their muskets had not been loaded. Lessard had never given the necessary order, and when it came to bluff Napoleon had the stronger nerves and the better cards. 'Everything is over,' he said to Drouot. 'In ten days we shall be in the Tuileries.'

While the Emperor made a speech to the new recruits and administered the oath to them, young Dumoulin came riding out from Grenoble dressed as a captain in the National Guard, wearing a tricolour cockade and carrying a promise of 100,000 francs from the Bonapartist faction in the town. It was turning into a day of victories, and as the light began to fail Napoleon felt sure that the fall of Grenoble would crown it. Lessard and his battalion were given the post of honour at the head of the column—a gesture which consolidated their new loyalties and would serve as an excellent example to their comrades in the garrison. And as they went down the steep descent into Vizille they met the 7th Regiment marching up to meet them, headed by its commander, Colonel Charles de la Bédoyère, who had produced the old regimental eagle and led his soldiers back to the man who had given it to them.

*

General Marchand had made one mistake after another, as if he were an indecisive politician rather than an experienced soldier. He had delayed too long before he sent the sappers to blow up the bridge at Ponthaut. He had left Lessard unsupported at La Mure. He had not even declared a state of siege to control civilian demonstrations in Grenoble, or done anything more than erect barricades in front of the main gates. But his worst mistake had been to fetch the 7th Regiment from Chambéry. Napoleon later claimed that Queen Hortense had somehow let him know that this regiment was seriously disaffected, and that Colonel de la Bédoyère—who had spent much of the winter in Paris and had been in touch with Hortense, Flahaut, and the dissident generals—was a brave and headstrong young man who might risk his head by opening the road to Paris. As early as 10 August in the previous year, in fact, Beugnot had asked an agent about de la Bédoyère's loyalty and been given an ambiguous answer, but the agent had added that 'his officers are in a very bad state of mind' and that both the *chasseurs* at Grenoble and the dragoons at Lyons were in a 'resentful' condition. For such reasons, Napoleon said, he had been anxious to know exactly where the 7th Regiment was stationed. Whether or not that claim was true the consequences of the regiment's defection at Grenoble were far more serious than they would have been at Chambéry, where it merely covered Napoleon's possible line of retreat into Switzerland, or over the Mont Cenis pass into Italy. And when Marchand let Charles de la Bédoyère and his men march out of the Porte de Bonne in a state of insurrectionary bravado, because he dare not call on the other units to detain them, he had lost his last chance of holding Grenoble.

It was then too late even to organize an orderly retreat. The crowds in the street were openly reading Napoleon's proclamations, and the soldiers were sporting tricolour cockades, chanting barrack-room jingles in favour of the Emperor, and calling over the walls to the Polish lancers and the crowd of excited peasants who had found another route to Grenoble when Lessard checked the main column at Laffrey. 'Will the officers fire if the men refuse to do so?' Marchand asked a royalist lieutenant of artillery. 'We should be hacked to pieces on our own guns if we did,' he replied.

At nine o'clock in the evening, when Napoleon and de la Bédoyère arrived, Marchand and the remnant of officers and men who remained loyal to the Bourbons were leaving through the Porte St Laurent on their way to Lyons. Campbell's prediction that Grenoble would be 'the test of Napoleon's success' had come true.

There was pandemonium at the Porte de Bonne, where the barricade had already been swept away. Outside in the torchlight was a shouting crowd of soldiers, peasants, and workmen. Marchand said they made a sound 'like the howling of wild beasts' when they began to assemble early in the evening and found that he had closed the city gates against them. On the walls were the equally excited men of the garrison regiments. The tension mounted when Napoleon went to the wicket of the Porte de Bonne and was rebuffed by Colonel Rousille, who had been holding the gateway for two hours with a small squad of reliable men. Then, as a group of wheelwrights began to pound away with a battering-ram, Rousille opened the door from within. It was a symbolic gesture which was greeted with a great shout, and as Napoleon was carried in shoulder-high people started to put candles in their windows. Since no one could find the keys of the town the artisans who had hammered at the great iron-bound door broke the wood into pieces, and came to lay them under the Emperor's window at the Hôtel des Trois Dauphins.

'Before Grenoble I was an adventurer,' Napoleon said at St Helena. 'At Grenoble I became a reigning prince again.'

20

CAUGHT IN A WHIRLWIND

Napoleon had covered 200 miles in six days, and the easy seizure of Grenoble was both a reward for the hard forced march over the mountains and an encouraging augury for what might happen all the way to Paris. He could afford to pause for thirty-six hours while the *grognards* rested and he played himself back into his imperial role.

The change was so simple and complete that the morning after Napoleon entered Grenoble it seemed as though the Bourbon regime had never existed. Emperor once more, at least so far as the Dauphiné was concerned, he wrote to Marie-Louise and to her father, announcing his return to France and asking her to resume her place at his side. He may not have had much hope of a positive reply, but it was a gesture he had to make for the sake of appearances, since he had been claiming all the way from the coast that the Austrians had known and approved in advance of his venture.

Emperor again, in Grenoble, he received the local dignitaries, charming them as he always did when he had something to gain by a show of interest and reasonableness, and reassuring those who feared recriminations for taking office under the Bourbons. 'The abdication had no binding force', he said in a phrase which characteristically fused his own dynastic claims with the rhetoric of the Revolution and the resentment against the way the Allies had reinstated the Bourbons, 'since it was never accepted by the people and it was imposed by foreign bayonets.'

Emperor indeed, among his soldiers, he received loyal addresses from those who had just joined him, some spontaneous, some stage-managed by Bertrand, for that master of protocol had now been assigned to organize the troops who defected, and to use them as an example to others who might be thinking of breaking their oaths to the King and rallying to the tricolour and the eagles. And before the fast-lengthening column moved off on the road to Lyons, the latest recruits at its head, Napoleon held a review in the Place de Grenette. This was the most familiar act of all, with the crowd singing the 'Marseillaise', the drums beating, and the veterans shouting 'Marengo!', 'Austerlitz!', 'Eylau!', 'Friedland!' and other battle honours as they paraded past him.

For the first time since the abdication almost a year before Napoleon had a proper force under his command. He could already count on 4,000 seasoned infantry, 20 roadworthy cannon, and a regiment of hussars, and discharged officers and men were streaming in all the time. If it was still too small a force to win a pitched battle, let alone take Lyons if the second city of France was conscientiously defended, the mood of the soldiery in Grenoble made such a battle seem increasingly unlikely. The civilians seemed infected by a revolutionary fever which ran through the countryside as fast as a man on a horse could carry it, and crossed over the Rhône into the streets of Lyons itself. When Napoleon left Grenoble on the following afternoon, riding in a carriage with only a few Polish lancers and staff officers to escort him, the journey was beginning to look more like a royal progress than a military adventure. There were sightseers everywhere; enthusiasts ran beside the cavalcade from one village to the next, and when Napoleon stopped for the night at Burgoin, the town was filled with people who danced and cheered as if they had just heard of a great victory. That was, in fact, exactly how it seemed to them.

★

The mood in Lyons was much the same. The city was so important that the King had despatched the Comte d'Artois, the duc d'Orléans, and Marshal Macdonald to hold it. But neither royalist zeal nor liberal sentiment nor military reputation could make up for the lack of reliable troops or persuade the recalcitrant citizens to make a stand for the failing Bourbon cause. .

The Comte d'Artois reached Lyons first, and there was no sign of the 30,000 men who were supposed to join him there. They were moving far too slowly to arrive in time. He could not even count on the garrison, though it was larger than the force which Napoleon had sent on from Grenoble and the city was in an easily defended position, for the soldiers were sullen to the point of open insubordination. When Artois turned them out for review on 8 March he gave each man a crown piece and still he failed to raise more than a few feeble cries of support for his royal brother. 'Those isolated cheers don't convince me,' the governor of Lyons said frankly to the prince. 'Look at that company of dragoons. They are positively scowling at you.' And though barricades were erected in front of the Morand and Guillotière bridges over the Rhône there were only two worn-out guns to support any resistance, and the local people seemed as keen to pull down the barricades as they were to stop the army engineers demolishing the bridges. The fact that not one bridge was damaged along Napoleon's whole line of march from Golfe Juan to Paris was just as telling as the refusal of any army unit to oppose him.

When the duc d'Orléans arrived next day he quickly decided that the situation was hopeless. 'It seems to me', he said to the Comte d'Artois, 'that the only thing you can do is to withdraw.' Marshal Macdonald, who arrived at nine o'clock in the evening and immediately held a council of war, came to much the same conclusion when the senior officers told him that their men would not fire on Napoleon and their comrades. All the same, Macdonald decided to make a last attempt to save the city. He was a dour, tactful, and trusted commander, the only one of Napoleon's marshals to be given his baton on the battlefield, and he hoped that the soldiers of the garrison might listen to him 'talk to them in their own language' at six o'clock next morning.

That night there were many meetings in taverns and barrack rooms. 'The officers are as excited as the men,' General Brayer reported before the parade on the Place Bellecour. 'As for me, I agree with them.' This was disturbing news, and although Macdonald himself was greeted with shouts of welcome the soldiers received his appeal in hostile silence. They were now very close to the borderline between duty and desertion, and when the Comte d'Artois came to walk through the ranks he almost pushed them too far by vainly trying to bully one dragoon into crying 'Vive le Roi!' Even this intemperate reactionary now saw that there was nothing to be done, and as he rode back to the archbishop's palace he could see the crowds gathering along the banks of the river in the hope of greeting the Emperor. The duc d'Orléans, always careful to avoid embarrassment, had already left for Paris, and the Comte d'Artois now followed him.

Marshal Macdonald, however, was a man who stood his ground obstinately: long ago he had lost the battle of the Trebbia to Marshal Suvarov because he refused to retreat. He went back to the barricade at the Guillotière bridge in the hope of finding a score of volunteers who would help him man it and thus provide a rallying-point for loyal troops, but the royalist mayor told him bluntly that there was no one in the whole city who would be willing to risk his

life for the King. As a last resort Macdonald decided to take a musket in his own hands and set an example. Even that proved to be
impossible. Before he could reach the barricade he saw that the
leading files of Napoleon's hussars, surrounded by jubilant peasants
and workers from the silk factories, had arrived first, and as they
tore down the pile of barrels and timber the soldiers from the garrison rushed forward to fraternize with them. It was the story of Laffrey and Grenoble all over again, and the excitement gave Macdonald a chance to mount his horse and ride away. A few of the
cavalrymen chased him right out on the road to the north, but
Macdonald was not to be caught. He was on his way to Paris to get
the King away to safety before the collapse he had seen in Lyons
was repeated in the capital.

The Emperor arrived about nine o'clock in the evening, and he
went directly to the apartment in the archbishop's palace which had
just been vacated by the Comte d'Artois. By then the city was in
the full swing of celebration, and the huge crowds which welcomed
Napoleon went on to maraud the streets in long torchlight processions, breaking the windows of known royalists and chanting slogans that soon moved on from the customary 'Vive l'Empereur' to
older and more sinister cries. 'Down with the Priests!' they
shouted. 'Hang the Aristocrats!' 'Death to the Bourbons!' A good
many people in Lyons that night remembered the scenes in 1793,
when the men of Marseilles marched up from the south singing the
great marching song of the Revolution.

★

Napoleon remembered too. He had always had ambivalent feelings
about the power of the people. In principle he appealed to it, and
larded his speeches with demagogic phrases. 'Liberty and the rights
of every Frenchman!' sounded uncomfortably familiar to moderate
ears. In practice, however, Napoleon sought to curb the notion of
popular sovereignty. He knew that his personal power rested upon
those who had profited from the Revolution yet were equally afraid
of losing their property to the Bourbons and their heads to the
Jacobins. He certainly had no desire to march on Paris—as Metternich put it in a graphic phrase—'like Robespierre on a horse'. And
so, in the three days he spent in Lyons, he made it quite clear that
he had come to restore the Empire, and order.

He began with a calming round of formalities, receiving the usual
deputations, collecting petitions, making public addresses and new

appointments. Fleury de Chaboulon had turned up again, and was made one of Napoleon's secretaries as a gesture of thanks for his mission; several old retainers and subordinates were also on their way south to join him. He made a conciliatory gesture by giving the Legion of Honour to a member of the National Guard who had found the courage to escort the Comte d'Artois out of the city. He set Bertrand to work organizing the soldiers from Elba, Grenoble, and Lyons into proper military formations, for he now had the makings of a far better army than the exhausted units he had deployed in front of Paris during the last weeks of the 1814 campaign. He had eleven regiments in all, seven of them seasoned infantry, two cavalry, one of artillery (and over fifty guns), and one of trained engineers. There were also half a dozen hotchpotch battalions, made up largely of half-pay officers and men who had returned to the colours as the column moved towards Lyons. Above all, to demonstrate that he had resumed the reins of power, Napoleon began to form a provisional government and issued a set of decrees which effectively dismantled the Bourbon system.

The first names on Napoleon's list were revealing. Marshal Davout, who had been away holding Hamburg to the last when the other marshals had forced the abdication at Fontainebleau, was to be Minister of War. The gourmandizing lawyer Jean Cambacérès, who had backed Bonaparte at the time he set up the Consulate in 1799, and drafted much of the Code Napoleon, was to be Minister of Justice. And there was, of course, to be a place for Fouché. In this crisis, and exactly as he had hoped, there was a splendid opportunity for this man for all intrigues. At the moment when Napoleon, in Lyons, was expecting him to be Minister of Police again, he was still in Paris and busy shuffling his contacts to find the most rewarding combination.

A trusted soldier. A resourceful politician. A cunning policeman. Napoleon thus symbolized the pillars of his state, and the decrees which followed filled the spaces in between. The Bourbon flag and cockade were suppressed; so were the old titles of nobility, the royalist orders of chivalry, all army and navy commissions granted to returning *émigrés*, all military promotions made in the past year, and all the nominations to the Legion of Honour which the Bourbon princes had scattered with such prodigality that it had been a prime cause of discontent in the army. The changes which the Bourbons had made in the courts of justice were withdrawn. Property which had been given back to its former royalist owners was again sequestered, while those who had acquired national lands during the Revolution and the Empire were confirmed in their possession. And to give a fresh and formal sanction to these reversals of fortune Napoleon proposed to call a curious assembly, called the Champ de Mai, to replace the Chamber of Deputies and the Chamber of Peers. This body, which took its name from the historic annual gatherings of the Franks, was to draft a new constitution and to provide the setting for the coronation of the Empress and the King of Rome.

The first of these notions could serve a useful turn. Napoleon knew that times had changed and that he must now woo the liberals and put a better constitutional gloss on his imperial system. But the second was bizarre, in all the circumstances, for it would soon turn out that the Emperor was unable to produce his wife and son for such a ceremony. It only made sense if he was doing everything he could to create an impression of confidence, partly to bluff the Bourbons off the throne ('my friends,' he said to his troops before they left Lyons, 'we are going to march to Paris with our hands in our pockets'), and partly to persuade the Allies that he truly intended to accept a peaceful part in the Concert of Europe.

★

It was already too late for that. Napoleon's attempt to pass himself off as a man of peace may have had some substance to it, though his record was bound to make any claim of this kind seem unconvincing; or it may have been nothing more than a tactic to confuse his enemies and delay any counter-measures until he had mobilized France for a new campaign. But his intentions were never put to the test. As soon as the news of his escape reached Vienna it was certain that, whatever Napoleon now said or did, Talleyrand and Metternich would seize their chance to deal with him once and for all.

The matter was really settled at the meeting which Metternich called as soon as he received the startling despatch from Genoa, and Talleyrand was then given the task of drafting a document which would justify a war and make another peace with Bonaparte impossible. 'The sovereigns of Europe', the declaration signed on 13 March insisted, 'would be ready to give the King of France and the French nation the assistance necessary to restore peace.' This seemingly innocuous phrase neatly transformed an act of rebellion against the King of France into a contingent act of war against the rest of Europe, and it was followed by an equally ingenious sentence which amounted to a bill of attainder. 'The Powers declare that, by breaking the agreement which had established him on the island of Elba, Napoleon has destroyed the sole legal title to which his existence was bound; that by reappearing in France he has placed himself beyond civic and social relations; and that, as an enemy and disturber of the peace of the world, he has delivered himself up to public vengeance.'

The claim that Napoleon had lost his 'sole legal right' to existence by flouting the Treaty of Fontainebleau was tactically clever but morally shaky, since Talleyrand and his colleagues in Vienna had been contemplating the Emperor's forcible removal from Elba for several months, and since the French had already ignored or deliberately violated every clause in that treaty. And as Britain had never signed the treaty in the first place Castlereagh found it difficult to persuade his critics in the House of Commons that Napoleon's return to France was in itself a reasonable cause for a new and expensive war against him. Talleyrand, however, did not really need to turn the treaty to such dubious account: that was merely an additional twist of spite. Once Napoleon left Elba the Powers were bound to treat him as 'an enemy and disturber of the peace of the world', whatever formal excuse they found to justify another invasion of France.

They had, in short, made him an outlaw, subject to summary justice if he should fall into their hands, and the signatures to the declaration read like a roll-call of judges on a warrant of execution. Metternich, for Austria; Talleyrand, for France; Wellington, for Great Britain; Palmella, for Portugal; Hardenberg, for Prussia; Razumovsky, for Russia; Labrador, for Spain; and Lowenhielm, for Sweden. They were all determined, they said, 'to guard themselves against every attempt which shall threaten to re-plunge the world in the disorders and miseries of revolution'. And to consolidate that purpose, on 18 March, they signed a new treaty of alliance in which all the parties pledged themselves to stay in the field as long as Napoleon was capable of further mischief—or, as a cynic might think after the experience of their previous coalitions, as long as Britain was willing to pay millions of pounds to subsidize their armies. The Low Countries then subscribed to the treaty; so did Sardinia, Hanover, Württemberg, Baden, Hesse, and Brunswick. Almost all of Europe was combined against Napoleon in the few days it would take him to travel from Lyons to Paris.

If he was able to complete that journey without hindrance from the royalist armies being assembled in Burgundy and the Bourbonnais there was no chance of peace. That was a sombre thought in a city that had spent the winter carousing towards a peace settlement. 'A thousand candles', wrote one of the men who had danced under them at every ball in Vienna, 'seemed to have been extinguished in a single moment.' And Lord Clancarty, who took over the British delegation when Wellington left to take up his command in Brussels, was distressed by signs of panic. 'It is not difficult', he said bluntly, 'to perceive that fear was predominant in all the Imperial and Royal personages.' Even when the odds against him were a hundred to one the spectre of Napoleon still haunted the throne-rooms of Europe.

★

The news reached Paris at noon on 5 March, when a message from Massena came up the line of semaphore stations from Lyons. Unlike the sovereigns in Vienna, who were pricked at once into vengeful anger, Louis received the report with the complacency which marked and vitiated everything he did. 'I am convinced', he told the foreign ambassadors on 7 March, 'that this event will have as little effect on the tranquillity of Europe as it has upon my mind.'

The ambitious and unprincipled Fouché, who heard the news almost as soon as the King, reacted very differently. Determined to turn Napoleon's sudden appearance to his own advantage, to ensure that whatever plot succeeded he would play a leading part in it, and to defend what he called 'the interests produced by twenty-five years of troubles', he began to intrigue more energetically than ever. No one liked him. No one trusted him. But his years of power, as he coyly remarked himself, had left him with so many 'links to public affairs', and 'so devoted a body of clients in the capital', that no one could do without him. Though he thought the Bourbons had ruined themselves, he offered to save them in case they managed to survive. Though he dismissed Napoleon as 'a worn-out actor, whose first performance could not be repeated', and insisted that he had long sought for 'some half-way house, some form of compromise which might avoid the desperate expedient of the Emperor's return', he made it clear to Hortense, Maret, and other loyal Bonapartists that he would be willing to serve the returning Emperor if he actually reached Paris. He let the Jacobins know that he remained a true friend of the patriotic party; he encouraged the

liberals; he kept one eye on the possibility of making the duc d'Orléans the saviour of France and the other on the prospect of doing
the same for Marie-Louise as regent for the King of Rome; and
though he was uneasy about the clique of dissident generals whose
restlessness made them incautious and incompetent plotters, he was
also party to their plan for a military rising and he hoped to be the
main beneficiary if it succeeded.

The details of this risky scheme remain obscure, for the relevant
records in the Ministry of Police seem to have been destroyed
(probably by Fouché) before the Bourbons returned to Paris in the
summer of 1815, and after that it was wise for anyone who had
been involved to keep its secrets. It is possible, moreover, that the
plotters had not completed their preparations when they were
caught out by Napoleon's unexpected return to France, and that
they were therefore obliged to improvise their response. But the
overall intention seems clear. The first phase was to be purely military. General Drouet d'Erlon was to raise the troops of the 16th
District, which covered the Pas de Calais and the Belgian frontier,
thus cutting the escape routes to the Channel ports and Brussels,
obstructing any attempt at intervention by the British units
stationed in Belgium, and providing a base for a march on Paris
itself. As that march proceeded, and more men were recruited from
the garrisons along the way, the second phase would begin. Half-pay
officers were to rouse the faubourgs, taking some of the Paris mob
to meet the oncoming army as it came into Paris through St Denis,
and others to the Tuileries to seize the King and other members of the royal household as hostages. At that point—at least in
the version of the drama which Fouché had written for himself—
the revolt would enter its third and final phase. Fouché would form
a provisional government, summon the Chamber of Deputies to
give it constitutional endorsement, turn out the National Guard to
support it, and then decide, according to the balance of forces,
whether to make France a republic again, or to set up an Orleanist
monarchy, or even to call Napoleon back from Elba.

This rising, it seems certain, was planned without Napoleon's
knowledge, and without any warning of his intentions, and his
return drastically reduced the range of options. Yet even with the
Emperor back on French soil the gamble was still worth taking.
Fouché, indeed, must have felt that the Emperor's unheralded
appearance was a good reason to strike at once, even if the preparations for the revolt were incomplete. For if the plotters triumphed

they might be able to bargain with Napoleon and impose limits on his power before he entered Paris, and if he succeeded they could at least claim credit for helping him to turn Louis off the throne.

As soon as Fouché learnt that Napoleon had landed at Golfe Juan, therefore, he sent a messenger off to Lille to inform the plotters. They reacted impulsively, and to little effect. Drouet d'Erlon's attempt to lead the Lille garrison towards Paris was thwarted as soon as it started, for his superior officer, Marshal Mortier, returned unexpectedly, contradicted the tale that there had been a revolt in the capital, and sent the troops back to their quarters. But by then the three other generals had their men on the move. Lefebvre-Desnoëttes, whose wife was the natural daughter of Joseph Bonaparte, was an unrepentant patriot, and the Lallemand brothers shared his nostalgia for the years of revolution and empire, and when they left Cambrai on 9 March with a small force they went first to La Fère in the hope of taking the arsenal and acquiring a stock of arms to equip their supporters. They managed to rally a regiment of *chasseurs*, some dragoons and a troop of gunners, but they were not strong enough to take the arsenal or persuasive enough to suborn its defenders, and they were turned away. They had no better luck next day, when they tried to take over an artillery park at Laon, and for the next three days they wandered disconsolately about the countryside between Paris and the Belgian frontier. The whole affair seemed more like a military exercise than a revolt, and when the column reached Compiègne without attracting any recruits, and Lefebvre-Desnoëttes suggested that they should save themselves by marching south to meet Napoleon, morale was so low that the majority preferred to surrender.

Drouet d'Erlon was already in prison at Lille. The Lallemand brothers were arrested and taken to Soissons, and Lefebvre-Desnoëttes and some other officers went into hiding. Only Colonel Marin, an artilleryman, managed to get away to join Napoleon at Auxerre a week later, and he received an ambivalent welcome. As Napoleon listened to Marin's account of the abortive revolt he seemed oddly pleased, as if it were proof that neither the army nor France could be won without him, and he also had a shrewd idea of the truth. 'The leaders of the conspiracy would have liked to take over the business and work for themselves,' he caustically told Fleury de Chaboulon. 'They claim to have tried to open the way to Paris for me. I know now what to believe. It is the nation, the people, the soldiers and warrant-officers who have done it all. I owe everything to them.'

The news that the revolt had fizzled out reached Paris much more quickly, and Fouché shrugged off that disastrous venture as casually as he had prompted it. Three days after the plotters fled at Compiègne he was involved in secret negotiations with the royalists.

★

For the moment, then, there were still two kings in France, and the royalist manifesto which was posted in Paris on 12 March put the situation in stark terms. 'France', it declared, 'will never be defeated in this struggle of liberty against tyranny, of loyalty against treason, of Louis XVIII against Bonaparte!' And to drive the point home Louis set about forming field armies to crush the rebellion, called for the raising of nearly three million National Guards to preserve order and provide reserves, and decreed that anyone who recruited troops for Napoleon would be liable to the death penalty. A few days later, as the news that one regiment after another was defecting filtered into Paris, a Bonapartist wag made the neatest reply to this wordy show of force. 'From Napoleon to Louis XVIII,' ran a placard hung on the railings of the victory column in the Place Vendôme: 'My good brother, there is no need to send me any more troops. I have enough.'

By the time Napoleon left Lyons on 13 March, in fact, all the country to the north of him was seized by the developing chaos which comes with a military collapse, when every item of news is confused, when no one knows what to do, when everyone gives orders and no one obeys them. An army in such a state is like a tattered flag on which the old emblems survive without the power to make men follow them. But in this case the cause of disorganization was not defeat by an enemy. The breakdown of discipline was due to a heady sense of release. The French army was full of soldiers who had felt like prisoners of the Bourbons in their own barracks, and now—as Napoleon had put it in his proclamation at Golfe Juan—saw themselves as 'the liberators of the motherland'. The march on Paris, Castlereagh explained to the House of Commons on 7 April, 'was a revolution effected by the army—effected by artifice—and by the sort of overweening influence which a person long at the head of a military system, and addressing himself to great military bodies, may be supposed to have possessed and exerted'. But Castlereagh saw only part of the truth about this remarkable event. It was actually one of those rare occasions in history when an army finds itself wholly at one with popular feeling and at once a surge of euphoria makes everything in the world seem possible. For the mass of the soldiers, and for the peasants and artisans from whom they were recruited, the return of Napoleon was a kind of second coming which would again illuminate France with the glow of glory and fraternity they had known in their youth.

The revolutionary temper rose fastest wherever the Emperor himself appeared. At Villefranche, where he paused after leaving Lyons on 13 March, there were fifty thousand people gathered to cheer him, despite the cold drizzle. There was a tree of liberty in the square, and tricolours everywhere, and eagles of gilt paper in the windows. There was a similar crowd at Mâcon that night, at Tournus next day, and again at Châlon on the following night. On 15 March, when Napoleon stopped at Autun, the townspeople were burning royalist flags and posters.

But it did not need Napoleon in person to raise a crowd. In many places the news that he really had landed in France and was marching northwards was sufficient encouragement. An insurrectionary mob seized control in Dijon. The same thing happened in Dôle, in Beaune, and in a hundred villages of Burgundy and the Franche-Comté, which had suffered a good deal under the recent Austrian occupation. 'In the name of the Emperor,' the retired General Allix cried on the square at Clamecy on 15 March, 'I assume command of the town.' That was all that was needed. On the same day there was a bloodless rising in Nevers, and at Auxonne, where Napoleon had served as a young lieutenant, the soldiers took over the artillery depot as the people took over the streets. In the chain of towns along the Saône and the Loire, where the two main roads ran towards Paris and Napoleon might soon pass, officials began to fetch out busts of the Emperor and Marie-Louise which had prudently been stored in municipal attics since the Restoration; and the ordinary people, so much less calculating in their response, made sure that the tricolour flew from one steeple to the next.

All over France, indeed, the authorities were reporting 'criminal acts' as royalist proclamations were torn down, and even the National Guard refused to do more than maintain order. 'I am afraid of civil war,' the Prefect of the Hérault wrote. 'The lower classes are very badly disposed', declared the Prefect in Rouen, five hundred miles to the north; and Marshal Jourdan, who was in command there, added an ominous rider to his claim that his troops were still calm. 'I don't know what they would do', he remarked, 'if their loyalty were put to any serious test.' In almost every garrison the commandants and the police agents were reaching the same conclusion. The men were either mute with hostility towards the Bourbons or they were openly mutinous, throwing away their white cockades and shouting for the Emperor; many army units near his expected route set out to join him. Even as Napoleon left Lyons the 3rd Hussars, the 23rd, 36th, 39th, 72nd, and 76th Infantry, all stationed ahead of him between Moulins and Bourg, brought out their hidden eagles, and he knew that both the Bourbonnais and Burgundy routes to Paris were open.

He chose to go by the more easterly route through Burgundy, over the high heathlands of Morven where the late snow still lay in patches, and down to the valley of the Yonne. As he covered stages of fifty miles or more a day, he kept overtaking units which had recently defected and had been turned round by General Girard, who had taken over command of the advance guard from Cambronne. He could not go any faster or he would reach fresh units before the proselytizing orderly officers riding on ahead had time to demoralize and convert them—a couple of these officers were arrested when they arrived too soon and met commanders who were loyal to the King. And he could not hold back in case the momentum of his dash across France was lost.

It was clearly impossible for the original column to keep up with him. The *grognards* from Portoferraio, for instance, only marched as far as Châlon, where they and some three thousand other soldiers were loaded on to river-boats to sail for at least part of the way on the Saône. By the time Napoleon reached Avallon on 16 March the infantry divisions and the artillery were strung out over a hundred miles of bad road, and they were in no condition to fight a battle. The whole march was in such a state of disorganization that, apart from the surging morale of the troops, it looked more like a retreat, and a resolute opponent—with a couple of reliable regiments— could easily have checked Napoleon at this point or even cut into his improvised army and captured him.

That task, in fact, had been assigned to Marshal Ney, who had left Paris swearing that he would carry Napoleon back to the capital in an iron cage. A year ago he had played the leading role in the abdication drama at Fontainebleau. Now the Emperor's fate lay in his hands for the second time.

★

Ney had much in common with Murat. He was a brave but

temperamentally unstable cavalry commander who had fought his way up from very humble beginnings, a man of greater ambition than character, whose driving vanity was soon to be his downfall; and in this crisis the conflict between his new attachment to King Louis and his old loyalty to the Emperor Napoleon left him feeling anxious and confused. As he passed through Poligny on 11 March he told the sub-prefect that Napoleon was 'a mad dog, and one must get rid of him to avoid being bitten'; when he arrived at Lons-le-Saunier, where his corps had been ordered to concentrate, he was still full of royalist bombast. 'If I can assure the triumph of the King,' he declared, 'I shall be the liberator of my country.' That prospect began to fade as Ney learnt that the situation was deteriorating all over Burgundy, and his own troops seemed reluctant to dash towards Mâcon and intercept Napoleon. All the same he struck a defiant posture and proposed to set his men an example. 'Give me a musket', he boasted, 'and I will fire the first shot.'

On 12 March, however, a royalist in flight from Lyons brought Ney a copy of Napoleon's proclamation to the army, and he was much affected by this reminder of past campaigns. 'The King ought to write like that,' he said, repeating the sonorous phrases aloud. 'That's the way to write for soldiers—that gets them!' And he had begun to speculate about his personal position if Napoleon should triumph. 'That madman will never forgive me for the abdication', he cried with the morbid elation which seemed to govern his behaviour during these exciting days. Yet that particular fear was allayed sometime during the night when two old comrades turned up on a clandestine mission from the Emperor, for they brought a personal note in which he promised that Ney would once again be welcome at his side; and they added their own persuasions to a long message from Bertrand which warned that any resistance might be the signal for a destructive civil war.

Ney was in a real muddle. 'I was caught in a whirlwind and I lost my head', he said candidly at the end of 1815, when he was on trial for his life. Even then he seemed uncertain about his motives, and unable to distinguish between his own knowledge of events and the half-truths and rumours which Napoleon's messengers had used to win him over. He talked about the growing discontent of the marshals, and the way the old aristocracy had snubbed their wives. He claimed that they had plotted to put the duc d' Orléans on the throne, and that the plot had collapsed when the duke had refused to play their game. He asserted that he had known that Hortense and other Bonapartists were intriguing for the Emperor's return, that General Köller had visited him at Elba and given an assurance of Austrian support if he would behave moderately in future, and that the British navy had connived at the escape from Portoferraio. That was an astonishing farrago of stale gossip for a man who had held high rank under Louis and Napoleon, and it was really no more than a repetition of the rambling speech Ney had delivered to General Lecourbe and General Bourmont on the morning of 14 March, when he warned them that he was about to defect.

The truth was probably much more pressing and much simpler. Ney had now realized that he could no longer control his soldiers. It was two days since the 76th Regiment, stationed at Bourg, had got rid of its officers, hoisted the tricolour, pillaged the wine rations, and marched off to meet Napoleon on his way from Châlon, and the other regiments in his corps were clearly on the point of mutiny. 'Can I hold back the waters of the sea with my hands?' he asked with a touch of pathos. It was essentially the same question that had faced Lessard at Laffrey, Marchand at Grenoble, and Macdonald at Lyons. Ney could join the Emperor, or he could run away from him, but he could not count upon a single company of soldiers to fight him. There was no incantation in the royalist repertoire which was powerful enough to dispel the imperial magic.

In the early afternoon of 14 March Ney formed his four remaining regiments into a hollow square. 'At that moment I looked at the soldiers,' one eye-witness said later. 'They all seemed gloomy and pale. I feared the return of one of those days in the Revolution when the officers fell victim to their men.' But in a few words Ney transformed the mood of mutiny into one of exaltation. 'The cause of the Bourbons is lost for ever', he said in a clear and steady voice. There was a great shout of 'Vive l'Empereur!' all round the square, and the men began to cheer wildly. 'Soldiers! I have often led you to victory,' Ney went on in the style that Napoleon himself used so effectively: 'now I am leading you to join the immortal phalanx that follows the Emperor to Paris!' While the officers who stayed loyal to their oaths withdrew, and one broke his sword in public, Ney 'went through the ranks like one possessed, embracing even the drummers and the fifers'. Next day he left to meet Napoleon at Auxerre.

★

Napoleon heard of Ney's defection before he left Autun, and the news was welcome for he was not in the best of spirits. He was running a heavy cold. He was hoarse from making impromptu speeches. He was tired by the journey. He was worried about the royalist army which deserting officers said was forming at Melun, outside Paris; he was afraid that he might yet have to fight for the capital, or that the King might withdraw to Tours or Bordeaux and force him into a drawn-out civil war; and in the meantime he was annoyed that all the notabilities and officials seemed to be scattering out of his way. 'The demonstrations of joy verged on delirium,' Fleury de Chaboulon noted as he watched the townspeople welcome Napoleon to Avallon. 'They squeezed, they suffocated, to get near him, to catch a glimpse of him, to speak to him.' But only one out of a dozen municipal councillors came forward to greet him. Napoleon needed the prefects, the mayors, and the police commissioners as symbols of order and continuity, to regulate the radical enthusiasm he had released in the population as a whole. 'I have come to forestall a terrible revolution', he told the mayor of Avallon, when that official had been fetched from his barricaded house to listen to one of the rambling speeches which Napoleon seemed to direct more at posterity than his immediate audience. 'I alone can spare France the evils with which she is threatened.'

The next day there was a sign of change. The Prefect of the Yonne was a banker called Gamot who was Ney's brother-in-law, and he had changed sides even more swiftly than the marshal. One day he was appealing for volunteers to join the King's army at Melun; on the next he was declaring that the Emperor alone 'can assure France the independence which will enable her to enjoy every kind of prosperity and give her the constitution appropriate to the character and customs of her people'. And though Napoleon later remarked contemptuously that 'Gamot behaved like a spaniel', on 17 March he was glad enough to see the prefect come riding out to Vermanton, where the imperial party had stopped for lunch, to say that an official welcome was being prepared at Auxerre. For the first time the Emperor was to be greeted in style.

Auxerre was an attractive and substantial town, slanting down to the wide span of the Yonne, and the party came rolling up to the bridge in the late afternoon. The lancers were smart in their turquoise uniforms with crimson facings, and the ostrich plumes nodding out of the odd squared-off hats the Poles call the *czapska*. Gamot and Drouot shared a carriage; so did Napoleon and Bertrand; and behind them came the imperial household. At the end of the bridge, in the crowd which almost blocked the way into the town, was the royalist mayor, Robinet de Malleville. 'You are come, Sire!' the mayor said portentously as he read from a formal address signed by all the royalist dignitaries who less than a week before had endorsed an enthusiastic declaration of loyalty to the King; 'and the national glory, inseparable from your own, is about to come into flower again.' The popular response was much more dramatic. There was a roar of welcome, a local doctor named Robineau Desvoidy remembered ten years afterwards. 'National cockade in hand, I too shouted with all the strength of my young lungs.'

Apart from Lyons, where Napoleon had occupied the archbishop's palace, he had been obliged to stay in post-houses and wayside inns, but in Auxerre, as protocol required, he was again installed in the prefecture. And there, as if to remind him of past grandeur, Gamot had ingratiatingly hung up a copy of Gerard's famous full-length portrait of the Emperor dressed in his coronation robes with his sceptre in his hand and a crown of laurel leaves on his brow.

The day had gone well, and it ended better. Before it was dark Napoleon was told that the 14th Regiment had marched in to the beat of the drum to herald Ney's approach. It was only four days since the same regiment had passed through Auxerre on its way south to catch Napoleon and put him in a cage. Now he came out to inspect the veterans who had fought under him for fifteen years, from Rivoli to Leipzig. He went on to meet the Abbé Viart, a contumacious priest, and picked the kind of quarrel he enjoyed about the pastoral letter in which the bishops had told Catholics to obey no one but the King; and then he returned to the prefecture, to give a celebration dinner. It was a special occasion, for the guests were Drouot, Bertrand, and Cambronne, who had shared his exile, and the three men who had won towns for him—de la Bédoyère, who gave him Grenoble; Brayer, who made sure that he entered Lyons without a fight; and Allix, who had taken over Clamecy all by himself.

Napoleon was in a relaxed and reminiscent mood that night, for he was among loyal friends, and for once he talked freely about the motives and the means of his escape from Elba. 'I am not in agreement with anyone,' he said, 'not even with those who are accused of conspiring on my behalf in Paris. From the island of Elba I saw the mistakes that were being made and I decided to profit from them. My enterprise has every appearance of an act of extraordinary audacity, but in reality it was simply an act of reason. There was no doubt that after all the harm that had been done to them the soldiers, the peasants and the middle-classes would welcome me with rapture.' There were two memorable phrases in that long evening's talk over the wine. 'I had only to knock on the door with my snuff-box and it would open,' Napoleon remarked. 'It was said last year that it was I who brought back the Bourbons. They are bringing me back this year. So we are now quits!'

21

IMPERIAL MAGIC

'I shall be in Paris when you receive this letter,' Napoleon wrote to Marie-Louise from Auxerre. 'Come and join me again with my son.' With only another hundred miles to go to the capital he had better reason to be confident about his own movements than about those of his wife and child, and while the perfunctory tone of his letters may be explained by his desire to avoid embarrassment should they be intercepted, they gave the same impression as those he wrote from Elba—they were the words of a man who was keeping up his appearances rather than his hopes.

Auxerre was to be the last stop of any importance before the dash to Paris, and Napoleon spent the morning of Saturday 16 March putting his affairs in some sort of order. He began by receiving Ney. The marshal had actually arrived on the previous evening and spent much of it drafting a self-justifying document, which Napoleon scarcely read and soon afterwards destroyed. 'My brave Ney has lost his wits', he remarked disparagingly, and he was equally caustic in a later comment to Hortense. 'Ney had the firm intention of attacking me', he said, 'but when he saw the troops he commanded resisting him he was simply forced to follow the movement and he sought to make a virtue of what he could not prevent.' Things might have gone differently for Napoleon that summer if he had kept to those sceptical opinions about his old subordinate. The real test of Ney's character, it turned out, was not the valiant command of the rearguard on the retreat from Moscow but the selfish panic of the abdication crisis a year before. And that still rankled, even though Napoleon found it politically convenient to accept Ney's act of contrition. The marshal was told there was no place for him on the triumphal march to Paris, and that he must be content with the mundane task of fetching the remainder of his troops from Dijon.

There were decisions to be made, and orders to be given, to organize the rest of the army that had so miraculously grown from the nucleus of two half-strength battalions that had sailed from Portoferraio just three weeks before. On paper it was now a most formidable force, which stretched back as far as Châlon and ahead through Joigny and Sens to Montereau: there, in the one brief but bloodless scuffle that occurred on the whole march, the 6th Lancers had seized the vital double bridge over the Yonne and the Seine which commanded the road through Melun and on to Paris. One way or another there were about 14,000 men marching under the tricolour, with several regiments of cavalry and a good many guns. The congestion on the road was so bad that Napoleon decided to put a good number of troops back on to boats, as he had done at Châlon, and float them down the Yonne.

Napoleon's staff, nevertheless, were not in a very festive mood. Fleury de Chaboulon had scraped together enough hints of a plot to assassinate the Emperor to revive the fears which had faded after he left Elba. Once again, as had been the case when Napoleon was going into exile, the blame was put on the Comte d'Artois, who was suspected of sending assassins to make an attempt as Napoleon passed through the Forest of Fontainebleau. Yet that was a minor worry compared to the difficulties which Drouot, Cambronne, Brayer, Girard, and the other commanders were encountering as they tried to find rations and fodder for an army which was driving across country without any previous planning, or proper lines of communication. They struggled to keep the formations closed up in case they had suddenly to be deployed for a battle, and sought to keep some measure of control over troops who were more jubilant than disciplined. The risk that the soldiers might get out of hand was reflected in Napoleon's order to General Girard when he heard that men in the advance guard were threatening reprisals against re-calcitrant royalists. 'Tell them', he said, 'that I would not wish to enter my capital at their head if their hands were stained with French blood.'

★

The situation in Paris was fast becoming more confused than the state of Napoleon's impromptu army. The King himself seemed unaware that his government was ceasing to function, and his ministers talked endlessly without coming to any useful conclusion. They had been deceiving the King and themselves for months, and now that the crisis was upon them they became victims of their own flattery, believing their own boasts about the royalist armies and tak- ing the sycophantic reports of royalist newspapers more seriously than the bad news which letter-carriers and travellers were bringing into the city every day. They found it very easy to assail Napoleon with verbal abuse. When it came to the point it was proving very hard to find anyone prepared to fire a bullet at him.

One of the few realists was Marshal Macdonald, who returned from Lyons on 15 March to warn the King that he should make plans to leave Paris since nothing could now stop Napoleon's advance. But no one in the Tuileries wanted to take Macdonald seriously. The ministers thought they might form a new government, get the King to win the liberals by a last-minute public conversion to constitutional principles, stir up fears of a new revolution among the middle-classes who were the backbone of the National Guard. Some of them were even prepared for the most sensational move of all. Perhaps, at the last, they could be saved by Fouché.

Blacas consulted him on 12 March, seemingly without effect. The next day he was seen by Chancellor Dambray, who was trying to put together a new government with the duc de Richelieu as its figurehead, the old revolutionary Carnot as its Minister of War, and Fouché as its moving spirit and Minister of Police. On 14 March Dambray went even further, and proposed that Fouché should lead the government in name as well as fact. That was a measure of desperation, and Fouché knew it; the days had been slipping away, and the royalists had left their bid too late for a man who could so well calculate the chances of success and failure. All the same, Fouché saw some advantage for the future in all these overtures, and he never neglected an opportunity to insure himself against changes of fortune. On the night of 15 March he went to the house of the Princesse de Vaudemont, a close friend who had been pressing his cause at the court, for a secret rendezvous with the Comte d'Artois.

It was a bizarre occasion, as the first regicide in France sat down with the first gentleman, the vengeful brother of the guillotined Louis XVI, to receive a proposal that he should form a liberal government, with full powers to deal with Napoleon and thereafter shape the domestic policies of France. He was flattered rather than persuaded. Nothing, Fouché insisted, could now prevent Napoleon's temporary triumph, and he admitted that he would serve the Emperor for the time being though he was convinced that the Allies would not tolerate Bonaparte and that they would soon restore the Bourbons for the second time. In that situation, he indicated, his services would again be available to smooth the transition from one regime to the other. As he neared sixty he had come to see himself as the spokesman for all those who wanted to stabilize French politics. 'Save the King', he is reputed to have told the Comte d'Artois, 'and I will save the monarchy.'

For the next few days, however, Fouché had to save himself. On the following morning, a police inspector arrived with a warrant to arrest him. It has been issued by Bourrienne, who had just taken charge of the police, and it was one of a score which named the leading Bonapartists in the capital. There were warrants out for Davout, Maret, Lavalette, Flahaut, and Excelmans, for instance, as well as for Fouché; and yet there was something odd about this apparently obvious move. Not one of these men was actually arrested, and it is not at all clear who had wanted them to be detained. Some people said that it was merely an act of personal spite by Bourrienne, who had been disgraced and dimissed by Napoleon years before. Others saw the hand of the King in the matter, and yet others, who knew Fouché's style, said that he had arranged this incompetent operation in order to cover his intrigues with the royalists and to present himself as a persecuted supporter of the Emperor. It is quite possible that the whole thing had been concocted when he had met the Comte d'Artois the night before, and he almost seemed to expect it. He coolly invited the inspector and his men to return to the house with him, left them waiting while he slipped out by a secret door, went up a ladder which had been conveniently set against the wall, down into the garden of Hortense, who lived next door, and away in a carriage which had been kept in her stables as a safeguard against such an emergency. For the next four days he hid in the Paris apartment of a former subordinate in the police ministry, and Hortense let all the Bonapartists know that Fouché had now thrown in his lot with them.

★

Strange times lead to strange and fleeting friendships. While Fouché went to ground the King went to the Palais Bourbon to meet the peers and deputies. In the end, Paris seemed worth a parliament. As he left the Tuileries he showed the duc d'Orléans that he was wearing the star of the Legion of Honour. 'I see it, Sire, but I would rather have seen it sooner', the duke remarked; it was a gesture that was as tardy and futile as the speech in which Louis, who had never shown any enthusiasm for the constitutional Charter, now hailed it as 'my proudest claim before posterity', and evoked an access of nervous enthusiasm from legislators who had no power and no means of rewarding the King's conversion to liberal principles. The most significant thing he said that afternoon, in fact, was a threat. 'The man who comes among us bringing the lighted brand of civil war', he declared, 'brings the scourge of foreign war as well.' If necessary, the Bourbons would once again prop themselves up with the bayonets of the Allied armies.

They clearly had little chance of being sustained by French muskets. As the King and the princes rode through Paris the crowd seemed sympathetic, but the soldiers who lined the route were in the sullen mood which had preceded outright mutiny all the way from Grenoble, and not even the distribution of money and spirits had roused them to cheer. 'Ah, how well they would fight,' the duc de Berry exclaimed with the vain optimism that characterized all the Bourbons. 'One or two may possibly make trouble, but they will be carried along with the rest.'

The royalist army, in truth, was little more than a figure of speech. The more distant garrisons were not too badly affected, as yet. Massena kept Marseilles for the King for almost three weeks, after he had prevented Pons from appealing to his troops by clapping that ebullient emissary into a cell in the Château d'If; the duc d'Angoulême in Bordeaux, the duc de Bourbon in Angers, and Marshal Augereau in Normandy could all keep their troops under control as long as they did not provoke any choice of loyalties. But to the south of Paris, where the real threat lay, there was no military order of any kind. The great army which was supposed to be gathering at Melun was nothing more than a demoralized muddle with units scattered randomly about the intended line of defence, and there was scarcely a regiment in which the officers had any definite orders or the men had any serious intention of obeying them; they were lucky if they could find rations and places to bivouac. When the duc de Berry came up to a group of grenadiers who were making themselves a meal somewhere near Villejuif he asked if he could taste the soup. It was a normal military courtesy, especially from Napoleon, but the Bourbon prince got a very sour answer. 'You will find it cold, sir,' one of the men replied. 'You have come too late.'

Back in Paris things were getting worse. On 17 March General Maison and General Dessolles went to see Blacas. They were indispensable men, for Maison commanded the garrison of Paris and Dessolles led the National Guard, yet they were not proposing to make themselves dispensable—unless they were each paid 200,000 francs to compensate them for the risks they were running in standing firm for the Bourbons. The news of Ney's defection had dramatically raised the market price of loyalty in Paris. And Blacas was foolish and desperate enough to pay this large sum within a few hours, for it was an afternoon of foolishness and desperation. He had already suggested that the King should lead the peers and deputies out to meet Napoleon near Fontainebleau, though it was not clear what he was supposed to do once they were face to face. Perhaps the devout Blacas may have hoped that Louis could repeat the miracle of Ste Geneviève, whose prayers were thought to have saved Paris from Attila the Hun.

Next day Louis issued a new proclamation to the army, which repeated the threat of approaching civil war and an ensuing Allied attack. It did nothing for morale, and worried those who were already frightened; it sent crowds to the passport office and the coach station, and pushed down the rate for the national funds on the Bourse. All over the more prosperous sections of Paris people were shutting up, selling up, packing up, and departing. In the course of five days the bankers converted over twenty million francs into foreign bills of exchange. The emigration was beginning all over again.

The following day was Palm Sunday. It was misty, and that added to the sense of uncertainty. There were rumours that the faubourgs were stirring but the only sign of change was the fact that the café owners had defied the sabbatarian law imposed on them months before and people had somewhere to go and shelter from the drizzle while they talked endlessly about the news. It was no longer a question of whether the King would leave for the north, only of when he would go, and as Macdonald went into the Tuileries in the middle of the day he saw carriages being loaded with royal baggage. It was better, he suggested, to keep them out of sight until darkness fell, for fear of exciting a riot. The last thing he wanted was to repeat the disaster of the flight to Varennes in 1791, when the royal family were caught and brought back to Paris. Anything could happen if there was too long an interval between the flight of the King and the arrival of the Emperor. Sixty miles to the south Napoleon had come to the same conclusion, and he sent messengers to tell his supporters in Paris to do all they could to keep the streets quiet.

There was, surprisingly, no trouble at all. Everyone in Paris seemed to be in a state of suspense, as if the city was caught in the dead eye of the storm that had so suddenly driven across France, sweeping one dynasty before it and bringing another in its train. There had been sightseers standing about the Tuileries earlier in the day, and they had seen the King go out for a short time to review his household troops in the Champs-Élysées, but by nightfall the squally rain had driven them all away. It was almost unnaturally quiet. 'Meanwhile, our death knell was sounding,' Vitrolles dramatically recalled. 'Our last moments had come, in the same indecision and inertia.' It was not even certain until seven in the evening that the King would take Macdonald's urgent advice and leave for Brussels, although Blacas seemed to take that decision for granted when he summoned the ministers to his house and gave 100,000 francs to each of those who promised to follow the King abroad.

All through the evening the windows of the Tuileries blazed with light, as if the occupants were attending a great party rather than packing their effects for a hurried departure. But the white Bourbon flag no longer fluttered over the palace. Someone said it had blown away, and whatever the reason for its disappearance no one bothered to replace it.

Then, towards midnight, the word spread that the King was about to leave. It was in his self-centred character that he did nothing to warn the royalists in the city whose lives or liberty might be most endangered by the return of the Bonapartes. It was all he could do to get himself away. Macdonald had thought of smuggling him out in a sedan-chair, but the fat and gouty monarch managed a slightly more dignified exit. He was a strange shape, well padded against the cold and, as a crowd of two hundred courtiers milled at the bottom of the staircase leading from the royal apartments, Blacas and the duc de Duras brought him out, half-helped, half-carried. While his last supporters wept openly he managed to utter a few regal and empty phrases. 'I shall see you again before long', he said consolingly, and then he was helped into his carriage, and he was gone. It was already Monday 20 March when the servants blew out the lamps in the courtyard.

★

Napoleon had spent all Sunday travelling, for he still feared that he might be forestalled by a popular rising, or a military coup, or some other combination which would prevent him entering Paris like a conquering hero. He lunched at Joigny. At five in the afternoon he passed through Sens, and it was long after dark when he stopped at the little town of Pont-sur-Yonne for dinner. Some of the men he had sent on in boats were there ahead of him, and he urged them to travel on through the night. In the confusion one boat was dashed against a bridge and thirty-three men of the 76th Regiment drowned. They were the only casualties of the whole journey.

Despite the disagreeable weather Napoleon kept moving. He left Pont-sur-Yonne at midnight, just as the King's carriage rolled away from the Tuileries. In the small hours he was at Fossard, where he had to decide whether to go straight on through Melun or turn away through the forest towards Fontainebleau and Essone. Still uncertain whether the duc de Berry had managed to hold some sort of force together at Melun, and confident that the 6th Lancers would protect his right flank by holding the Montereau bridge, he decided to go on to Fontainebleau before dawn. And there was another reason for his choice. Napoleon was fascinated by anniversaries and symbolic occasions and it seemed right to him to return to Fontainebleau, where he had begun his journey to Elba exactly eleven months before. At five o'clock in the morning a hundred cavalrymen escorted the Emperor into the Cheval Blanc courtyard of the palace. He was expected, for when he went straight to his apartment there was a bright fire burning and one of his former servants was there to show him that his crowned initial was still carved into the gilded wood-frame of his bed. He lay down half-dressed; for all the need for haste he had to rest. He had been on the road for twenty hours and he had forty miles yet to go if he was to celebrate his son's birthday in Paris.

That same morning, in Vienna, the Emperor Francis took the King of Rome away from his French attendants. He was determined that when he settled accounts with his son-in-law there would be no right of succession left for his grandson. From that day the inconvenient child ceased to be the heir to a dynasty and became something like a royal foundling.

*

All through Monday morning couriers were riding in to Fontainebleau with messages for the Emperor. They brought him despatches from his own commanders, as they moved up towards Paris; reports from units which had just defected, and had no idea what to do next; letters of welcome from Hortense, Maret, and other ardent Bonapartists; and most important of all, an urgent summons from

Lavalette, who had simply gone back to his old employment in the Post Office and at once sent word that the King was gone and that Napoleon should make for the Tuileries as soon as possible. That settled the matter. The Emperor would have to take the risk of moving into the capital with no more preparation than there was at Lyons. And once again it was Marshal Macdonald who faced him and realized that there was nothing to be done. The troops who had been expected to man a defence line at Villejuif were simply waiting for the Little Corporal to return to them.

The cavalcade actually made slow progress. Despite the showery weather there were crowds standing all along the muddied road, and as Napoleon's carriage forced its way through the people closed in behind it to form a rambling cheerful procession. It was six o'clock and the light was already fading when the Emperor reached Essonnes, where Marmont's corps had so ruinously gone over to the Austrians a year before, and where Napoleon now held a vast informal rendezvous of soldiers, peasants, clerks, shopkeepers, members of the National Guard, and hundreds of men, women, and children who had flocked to see this marvel of their age. All of them were strangely excited. If Napoleon had called upon them they would have marched with him to the walls of Paris as their fathers had gone to storm the Bastille and thrown themselves into the miraculous battles at Valmy and Jemappes in '92. But there was no sign of the men who had made their names in those stirring days. Where, indeed, were the marshals? Some, like Berthier, Macdonald, Marmont, and Mortier were keeping their royal oaths. While Ney had already been obliged to make his choice, others were waiting cautiously to see what happened before they changed one cockade for another. Of all the men whom Napoleon had honoured with titles and estates only Davout and the ageing, plain-spoken Lefebvre were to give him a spontaneous welcome. And though Caulaincourt, Flahaut, and some other cronies and aides came riding out to greet him on Monday afternoon, the men of place and power in Paris mostly stayed in their homes that day.

The one man who came out well from these trying times was the dour and honourable Macdonald. He had shown dignified courage at Lyons, realism in his assessment of the situation, and common sense in advising the King. His decision to leave the armies at Villejuif and Melun to their own devices and to carry the few loyal troops off to the north was the only sensible thing to do in the circumstances. The Comte de Lamothe-Langry, who spent most of

Monday calmly walking about Paris and noting what he saw, said that Macdonald's action 'forestalled all rivalry, all struggle; it averted a collision and prevented the shedding of any French blood'. What most worried the count during these anxious hours was the fact that the capital was left open to 'brigands and convicts ... pillage and arson'. Yet such fears proved groundless, and by the middle of the day the Bonapartists had begun to assume control.

General Sebastiani rode out to Villejuif, imposing sufficient discipline on the units which Macdonald had abandoned to begin moving them in to garrison the capital again; and in the north of the city General Excelmans suddenly implemented the last phase of the abortive military rising. He had gone out to St Denis in the hope that he could use disaffected soldiers to catch the King. When that scheme failed, and the King was safely on his way, Excelmans quickly raised a scratch force of half-pay officers, some cavalrymen, and a few gunners with a pair of cannon, and led them back to the centre of Paris. It was about two o'clock in the afternoon when he came clattering along towards the Tuileries, a huge tricolour bunting wound about him like a toga, and a few minutes later a Captain Mouras hoisted the tricolour over the royal palace. Soon afterwards the pensioners at the Invalides raised the red, white, and blue banner and fired a saluting cannon which made everyone in Paris think that Napoleon had arrived. Within the hour the national colours were flying everywhere in the capital, including Notre-Dame, and the eagle's flight was almost over.

★

Paris was astonishingly quiet. Many of the shops were closed as if it were a holiday, some as a precaution, some because their proprietors were away on duty in the National Guard. In the square of the Carrousel, which was the nearest people could get to the Tuileries, there were stalls set up to catch the unexpected trade, and there were women selling beer and brandy, and fried potatoes, sausages, snacks, and sweetmeats. But the crowd thinned out towards evening, for no one seemed to know what to expect and the mist was rolling up from the Seine. 'We had been so spoiled along the way that the welcome given the Emperor by the Parisians by no means came up to our expectations', Fleury de Chaboulon confessed. Perhaps Paris had always been cooler towards Napoleon than the towns and villages which had cheered him as he came up the Saône and down the Yonne. Perhaps the Parisians were more apprehensive and confused, for they had been told for the past week that the royalist army would crush the oncoming tyrant at the gates of the city—and now they had gone to bed with one government and risen with another. Perhaps, after all, Fleury was right, and the poor reception could be explained by lack of theatrical preparation. 'One must simply conclude', he said as an excuse, 'that Napoleon missed his entrance.'

But even if the ill-lit streets were almost deserted the Bonapartist faction had not missed its chance. For several hours Napoleon's supporters, and those who now wished to be counted among them, had been making their way to the Tuileries. As the domestic staff put everything back into imperial style—even ripping off the fleurs-de-lis which had been stitched on the furnishings to cover the Napoleonic emblems—the self-invited guests spilled up the stairways and into the great reception rooms. There were former ministers and councillors of state, generals, colonels, and majors, functionaries and policemen, all in uniform, with the red ribbon of the Legion of Honour in pride of place. There were women in court dress, and flunkeys in livery, and a scattering of men in cut-away coats and kerseymere breeches. And down below, in the courtyard, there were hundreds of officers milling about with suppressed excitement.

It was a little before nine o'clock when Napoleon and a small escort entered Paris through the Villejuif *barrière*, drove by the Invalides, crossed the Seine and turned into the Tuileries. 'Up to the gates, there being plenty of room, we travelled freely,' his valet recalled. 'But once we were inside the courtyard . . . it was impossible for me to drive the carriage up to the steps. The Emperor, seeing he could go no further, got out in the middle of the huge crowd which pressed around him.'

Like Louis on the previous evening Napoleon had to be half-helped, half-carried through the entrance, though for a very different reason. Louis was large, and he was small, and there was no other way to get him through the jostling throng and up the staircase. He had not tried to speak when people surged about the carriage. If he had done so, the cheering would have drowned his words. But as he was caught in the press of enthusiasm he made one cry of protest. 'My children, you are suffocàting me!' Caulaincourt immediately shouted at Lavalette, 'In God's name make a space for him', and Lavalette backed up along the balustrade to clear the way.

Dinner was served that night 'as if the Emperor had never left the Tuileries', and afterwards, keyed up by excitement and exhaustion, Napoleon walked among the guests for an hour to savour his triumph to the full. 'Everyone seemed drunk with happiness and hope,' his personal servant Marchand remembered. 'The Emperor himself was unable to hide his delight; I have never seen him so wild with glee, so lavish with pats on the cheek.' Only Drouot and Bertrand, who had shared his sudden fall and spectacular resurrection, could really tell what that moment meant to him. He seemed like a man bewitched, as if his translation from the Villa Mulini to the Tuileries had been an act of magic. And, in its own way, it was.

EPILOGUE

The reign of the King of Elba was an extraordinary and bizarre performance, like a harlequinade staged between two acts of a tragedy, and there has been nothing else quite like it. It was produced by a casual act of condescending generosity on the part of the Tsar, and it was brought to an end because the principal actor and his audience of statesmen had both tired of it.

Napoleon certainly made a change for the worse. After Waterloo he went back to Paris, and exactly a hundred days after his triumphant entry into the Tuileries he abdicated for the second time, fled to Rochefort in the hope of escaping to America, and then gave himself up to Captain Maitland on the *Bellerophon*. Once again he suggested that the Prince Regent should grant him asylum, and once again he was refused. All that he saw of England, in the end, was the sight of the Devon coast from Torbay and Plymouth Sound as he passed by on his way to the remote cliff-bound island of St Helena in the South Atlantic. After six years of exile he died on 5 May 1821 at the age of fifty-two.

The faithful Bertrand was at Napoleon's side at Waterloo, and took his wife Fanny and his children to St Helena, where they stayed until Napoleon's death; Bertrand then went home to play a modest role in politics, and serve as commandant of the École Polytechnique. In 1840 he was one of the group chosen to fetch Napoleon's ashes back to France, and he died in 1844. Drouot was also at Waterloo, in command of the Guard, and when the Bourbons came back and put him before a court martial he defended himself with dignity and was acquitted. So was Cambronne, picked up wounded and stripped by pillagers after leading the defiant last stand at Waterloo. Ney was less fortunate. He mishandled the battle of Quatre Bras, when he had Wellington at a disadvantage, and his mistakes at Waterloo two days later may well have cost Napoleon the battle. He too was arrested by the vengeful Bourbons, tried before a court of peers, and sent before a firing squad on 6 August 1815. Colonel de la Bédoyère met the same fate.

Colonel Campbell was on the other side at Waterloo, though

Castlereagh would not attach him to any of the Allied staffs. 'It was feared', Campbell sadly observed, 'that my presence might excite irritating discussions.' The best Castlereagh could do, as Campbell went to join his regiment in Flanders, was to send a letter assuring Wellington that the British Government 'had every reason to be satisfied with the activity and intelligence manifested by Sir Neil Campbell...during the very delicate and difficult task imposed upon him, while residing near the person of Napoleon Bonaparte'. Campbell fought well, and survived the battle, but the rumours of collusion in Napoleon's escape continued. In 1826, with the rank of major-general, he was sent to be Governor of Sierra Leone where he soon acquired a reputation for hard work and harsh judgement. He died of fever in the following year.

Murat was as unfortunate as Ney. Driven by ambition, and failing to understand what Napoleon wanted, he attacked the Austrians and tried to drive them back over the Alps; and since he lacked the resources and the skill for such a campaign he was soon defeated and driven to seek refuge in France. 'My brother-in-law has ruined me twice,' Napoleon said: 'the first time by deserting me, and the second time by supporting me.' Murat had also ruined himself. After Waterloo he copied Napoleon's landing at Golfe Juan in a reckless descent on the coast of Calabria, in the belief that the people of Naples would still prefer him to the Bourbons, but it came to nothing. He was captured, court-martialled, and shot within the hour.

Pons de l'Hérault had only a minor part in the Elban drama, although his recollections of it became a major source for its historians. Once he had resolved the conflict between his republican principles and personal admiration of Napoleon he remained devoted to the Emperor and even asked to share his exile on St Helena. He had a chequered career. Released by Massena after six weeks in the Château d'If he became Prefect of the Rhône, went into exile after Waterloo, returned to France in 1822, and lived to take part in the revolution of 1848 and to oppose the coronation of Napoleon III as a violation of republican faith. He was living in poverty when he died in 1858 at the age of seventy-nine. Peyrusse became mayor of his native Carcassonne in 1830, and survived to be honoured by Napoleon III before his death in 1860.

Napoleon's mother rightly feared that she would never see him again after he sailed from Portoferraio. A few days later she went to Rome, where she lived quietly until she died in 1836. Pauline crossed to Tuscany, where she was kept under house-arrest for some time. Her request to join her brother on St Helena was also refused. She died in Florence in 1825. Elisa spent her last years in Trieste where she befriended Fouché, exiled from France after stage-managing Napoleon's second abdication and the second restoration of Louis XVIII. She died five years after Waterloo.

Marie-Louise left her son in Vienna, where he was brought up as the Duke of Reichstadt. He was never allowed to correspond with his father, or to do anything which might remind people that he had once been a Bonaparte, and he led a cloistered and insignificant life until his early death at the age of twenty-one. Marie-Louise, meanwhile, had secretly married Neipperg, and they lived in Parma until his death in 1828. She then married again, this time to the Comte de Bombelles who had been one of the secret intermediaries in the negotiations between Francis and Louis in 1815. She died in 1847. Maria Walewska had a successful visit to Naples. On 30 November Murat revoked the decrees of confiscation on her son's endowment, paying over half a million francs in accumulated rents and interest. She played no part in the Hundred Days, and in 1816 she married again, this time to a cousin of Napoleon called General d'Ornano. She died in the following year. Alexander Walewski was said to be the image of his father, though much taller. He lived to become a French diplomat and to play a notable part in French public life.

That was not quite the end of the Bonapartes. Joseph, once King of Naples and King of Spain, managed to get away to the United States in 1815, and he lived comfortably in Philadelphia before he returned to die in Florence in 1844. Lucien was the only member of the family without dynastic ambitions, and he also made an attempt to settle in America, but he soon returned to Italy and lived in Rome until his death in 1840. Louis had been King of Holland for four years, and the reluctant husband of Hortense de Beauharnais, and after his brother came to grief he wandered from Germany to France to Switzerland before his death in Livorno in 1846. The youngest of the brothers was Jerome, who had been King of Westphalia. He had led the attack on the Hougoumont farm at Waterloo which had been a critical moment in the battle, and afterwards he moved about Europe without much purpose until he was allowed to return to France in 1847. The times had changed, and his nephew Louis Bonaparte was soon to become president and to make him a marshal; and in 1852, when President

Bonaparte proclaimed himself Emperor, and took the title of Napoleon III, Jerome became president of the Senate. He had lived long enough to see his brother's memory become a national myth, and the Bourbons remembered largely by the striking phrase from the Golfe Juan manifesto: 'They learned nothing and they forgot nothing.'

For the Bourbons did not last very long when they returned in the baggage-waggons of Wellington's army. Louis XVIII survived until 1824, a model of a sick, weak, and reactionary monarch. His brother, the Comte d'Artois, succeeded him as Charles X. In 1830 the duc d'Orléans was given his chance to be king of the French. As Louis Philippe, at last fulfilling Napoleon's graphic prediction, he became 'a sort of compromise between the Revolution and the Restoration'.

That might have been Napoleon's own destiny if he had been able to put his countrymen before his country, and preferred true greatness to the costly search for glory.

BIBLIOGRAPHICAL NOTE

The references given below are normally to the editions used in writing this book, but in the case of some translated works I have gone back to the original text to make a fresh and hopefully more comprehensible translation. Most readers will know something of the historical background to this book, since there are many biographies of Napoleon and these are easily accessible in libraries and bookshops. Those who wish to refresh their memories, and do not wish to consult a monumental work such as L. Madelin's *History of the Consulate and Empire*, Volumes XIV, XV, and XVI, will find J.M. Thompson, *The Life of Napoleon* (Oxford, 1952), G. Lefebvre, *Napoleon* (London, 1969), and Vincent Cronin, *Napoleon* (London, 1971) reliable standard works in English. J.C. Herold, *The Age of Napoleon* (London, 1963) and George Rudé, *Revolutionary Europe 1783–1815* (London, 1967) provide excellent contextual reading.

Memoirs

The participants in the events of 1814 and 1815 were prolific in memoirs, written by themselves or at least attributed to them, and they are full of interest if uncertainly reliable. The most relevant sources are given here in alphabetical order. G. Bertrand, *Les Cahiers de Sainte-Hélène* (Paris, 1949) records Napoleon's rambling recollections during his second exile. Louis de Bourrienne, *Mémoires sur Napoléon* was the work of a discredited secretary who became an active opponent; it is not trustworthy in the absence of corroborating evidence, but Volumes 9 and 10 effectively evoke contemporary attitudes. Armand de Caulaincourt's *Memoirs* (London, 1950) is the most comprehensive account of Napoleon's agonized fortnight at Fontainebleau in April 1814, and it is generally reliable on the events leading to his abdication. Napoleon's secretary, Baron Fain, wrote three important books: *Manuscrit de 1812* (Paris, 1827), *Manuscrit de 1813* (Paris, 1825), and *Manuscrit de 1814* (Paris, 1823). Fleury de Chaboulon wrote his *Memoirs* (London, 1819) while Napoleon was still alive and able to comment sarcastically on his self-important claims. Vague about dates and details, it is regrettably the only and unreliable source for some aspects of Fleury's visit to Elba. Joseph Fouché, *Mémoires* (London, 1825) is so clearly a synthetic work that it cannot be trusted, and it is frustratingly silent on many points on which the cleverest policeman in Europe undoubtedly knew the hidden truth. G. Gourgaud, *Journal* (Paris, 1945) is the work of a dedicated officer who fought in the 1814 campaign and later went with Napoleon to St Helena. J. Hanoteau edited the *Mémoires de la*

284 The Escape from Elba

Reine Hortense (Paris, 1927). E.P.D. Las Cases, who also went to St Helena, wrote Memoirs of the Emperor Napoleon and Le Mémorial de Sainte-Hélène (Paris, 1951) which are based upon Napoleon's own reminiscences. C. Rousset edited The Recollections of Marshal Macdonald (London, 1893). Marshal Marmont, Mémoires (Paris, 1856–7) and L. Marchand, Mémoires (Paris, 1952, 1955) are reasonably authentic, and contain more useful material than either the Memoirs of the valet Constant or the Souvenirs of Saint-Denis, known as Ali the Mameluke (Paris, 1926). C.F. Méneval, secretary to Marie-Louise, was an honest man, and his book Napoléon et Marie-Louise (Paris, 1844) was supplemented by his Mémoires (Paris, 1894). The Memoirs of Prince Metternich was edited by his son (New York, 1881). C.J.F.T. de Montholon was another of those who wrote down Napoleon's talk at St Helena, but his Mémoire de Napoléon is open to serious question, and there is even more evidence of serious plagiarism in his Récits (Paris, 1847). Hyde de Neuville, Mémoires et Souvenirs (Paris, 1894) seems as reliable as one may expect from a man who spent much of his life in clandestine activity. Guillaume Peyrusse published his Mémorial et Archive (Carcassonne, 1869), and L.G. Pelissier collected his Lettres inédites (Paris, 1894). General Savary, who became duc de Rovigo and succeeded Fouché as Minister of Police in 1810, wrote a History of the Emperor Napoleon (London, 1828) and a Mémoire sur l'Empire (Paris, 1821); like many of those writing memoirs soon after the Bourbon restoration, he made sure his text was acceptable by omissions and the expression of suitable opinions. P. de Ségur was another reminiscent general who let his adhesion to the Bourbons colour his Histoire et mémoire (Paris, 1857). Talleyrand was no more likely to tell the truth than Fouché, but the Mémoires (New York, 1891) that were edited by the duc de Broglie were at least based on an authentic text.

Biographies

Among the many biographies which relate to the Elban episode are the following recent or particularly relevant texts, which are grouped by subject. E.M. Almedingen, The Emperor Alexander I (New York, 1964)); M. Grunwald, Alexander I (Paris, 1955); Marjorie Weiner, The Sovereign Remedy (London, 1971). Walter Scott, Life of Napoleon Buonaparte (Edinburgh, 1827) was written close to the events described, and is remarkably well informed. F. Masson, Napoléon et sa famille (Paris, 1894), Monica Stirling, A Pride of Lions (London, 1961), R.F. Delderfield, The Golden Millstones (London, 1964), and David Staction, The Bonapartes (London, 1967) all deal with the Bonaparte family. Baron Larrey, Madame Mère (Paris, 1892), Alain Decaux, Napoleon's Mother (London, 1962) and G. Martineau, Madame Mère (London, 1977) are all biographies of Letizia Bonaparte. Bernard Nabonne, Pauline Bonaparte (Paris, 1948), Pierson Dixon, Pauline: Napoleon's Favourite Sister (London, 1964), and Len Ortzen, Imperial Venus (London, 1974), cover the available material on Pauline's two visits to Elba. Imbert Saint-Amand, Marie-Louise, l'île *d'Elbe, et les Cent Jours* (Paris, 1885) is a standard source, though it lacks more recently published material including the correspondence between Napoleon and his wife. Paul Garnière, *Corvisart, Médecin de Napoleon* (Paris, 1951) contains a good deal of information about the Empress, especially her sojourns at Blois, Rambouillet, and Aix, and Patrick Turnbull, *Napoleon's Second Empress* (London, 1971) is a useful general biography. Christine Sutherland, *Maria Walewska* (London, 1979) uses fresh material from Polish sources.

On other members of the family see: T. Iung, *Lucien Bonaparte et ses mémoires* (Paris, 1883), Joan Beer, *Caroline Murat* (London, 1972), M.H. Weil, *Joachim Murat* (Paris, 1909, 1910), especially the last two volumes, Paul Garnière, *Murat, roi de Naples* (Paris, 1959), and Carola Oman, *Napoleon's Viceroy* (London, 1966), which is a straightforward biography of Eugene Beauharnais. Louis Madelin wrote standard biographies of *Talleyrand* (Paris, 1944) and Fouché (1946). See also J.F. Bernard, *Talleyrand* (London, 1973), and Emile Dard, *Talleyrand and Napoleon* (London, 1937).

On the Allied statesmen see Charles Petrie, *Lord Liverpool and His Times* (London, 1954), J.A.R. Marriott, *Castlereagh* (London, 1936), and C.J. Bartlett, *Castlereagh* (London, 1966), G. de Bertier de Sauvigny, *Metternich et ses temps* (Paris, 1959), and D.G. McGuigan, *Metternich and the Duchess* (New York, 1977).

On the soldiers who accompanied Napoleon to Elba see: Jules Nollet, *Biographie de Génèral Drouot* (Paris, 1850), W. Serieyx, *Drouot et Napoléon* (Paris, 1929), L. Brunschvigg, *Cambronne* (Nantes, 1894), and Louis Garros, *Le général Cambronne* (Paris, 1949). Robert Christophe, *Les amours et les guerres de Maréchal Marmont* (Paris, 1955) is a fairly recent biography of the man whose havering led to the defection of the Sixth Corps. E.A. Vizetelly, *The Wild Marquis* (London, 1905) and Maurice Garçon, *La Tumultueuse existence de Maubreuil, Marquis d'Orvault* (Paris, 1954) discuss the mysterious Maubreuil affair in detail.

Politics

Henri Houssaye *1814* (Paris, 1882) remains a standard text on the events of that year. Edouard Driault, *La chute de l'Empire* (Paris, 1927), Jean Thiry, *Le premier abdication de Napoléon Ier* (Paris, 1939), Felix Ponteil, *La chute de Napoléon Ier*, and R.F. Delderfield, *Imperial Sunset* (London, 1968) all describe the collapse of the Empire in some detail. Charles Dupuis, *le Ministère de Talleyrand en 1814* (Paris, 1919), and G. de Bertier de Sauvigny, *The Bourbon Restoration* (Philadelphia, 1966) both deal with the formation of the Provisional Government and the return of Louis XVIII. There are numerous contemporary accounts by visitors to France after the fall of Napoleon, but the social background is well summarized in Simona Pakenham, *In the Absence of the Emperor: London-Paris 1814– 1815* (London, 1968). For the background of the escape and the march on Paris see Henri Houssaye, *Le retour de Napoléon* (Paris, 1934), Claude

Manceron, *Napoleon Recaptures Paris* (London, 1968), and P. Ravignant, *Le retour de l'île d'Elbe* (Paris, 1977), the last of which is an example of the popular-heroic mode of writing on this theme. Emile le Gallo, *Les Cent Jours* (Paris, 1924) is helpful on the peculiar conspiracy of the generals in the Nord. Among the curiosities are A.D.B. Monnier, *Une année de la vie de l'Empereur Napoléon* (Paris, 1815), C. Doris, *Secret Memoirs of Napoleon Bonaparte* (London, 1815) and *La Vérité sur les Cent Jours, Par un Citoyen de la Corse* (Brussels, 1825) which contains credible material on the Italian nationalist conspiracy. A delightful commentary on the Elban episode as portrayed by contemporary cartoonists is to be found in A.M. Broadley, *Napoleon in Caricature* (London, 1911).

The Congress of Vienna

The two standard works by Charles Webster are *The Foreign Policy of Castlereagh 1812–1815* (London, 1931) and *The Congress of Vienna* (London, 1946). Harold Nicolson, *The Congress of Vienna* (London, 1946) is a conventional and convenient summary of the issues. Frederick Freksa, *A Peace Congress of Intrigue* (New York, 1919) is based on the memoirs of participants, and three first-hand accounts can be found in Jean-Gabriel Eynard, *Journal* (Paris, 1914), Carl Bertuch, *Tagebuch vom Wiener Kongress*, edited by H. von Egloffstein (Berlin, 1916), and Niels Ronsenkrantz, *Journal du Congrès de Vienne* (Copenhagen, 1953). A most important source is M.H. Weil, 'Le Revirement de la Politique Autrichienne' (Turin, 1908). In this paper Commandant Weil reprinted the clandestine correspondence between the Emperor Francis and Louis XVIII, in which the two sovereigns were planning the overthrow of Murat in the weeks before Napoleon's escape.

Military

J.H. Rose, *Napoleon's Last Voyages* (London, 1906) includes the first-hand account of the crossing to Elba on the *Undaunted*. For books discussing Napoleon's relations with his former comrades-in-arms see: Marcel Dupont, *Napoléon et la trahison des maréchaux* (Paris, 1939), Louis Chardigny, *Les Maréchaux de Napoléon* (Paris, 1946), A.G. Macdonnell, *Napoleon and his Marshals* (London, 1950), and Peter Young, *Napoleon's Marshals* (London, 1974). Marcel Dupont *Napoléon et ses grognards* (Paris, 1945) is a useful source on the veterans. J.C. Lawford, *Napoleon's Last Campaigns* (London, 1977) and F.L. Petre, *Napoleon at Bay* (London, 1977) both describe the final stages of the long war. See also Jean Thiry, *La campagne de France de 1814* (Paris, 1946). Two valuable sources on military orders and other documents are Arthur Chuquet, *Lettres de 1814* and *Lettres de 1815* (Paris, 1914). For contemporary accounts of the campaign of 1815 see Anthony Brett-James, *The Hundred Days* (London, 1964) and for an overall summary of the events after Napoleon's return see Edith Saunders, *The Hundred Days* (London, 1964).

Police

Material on police and espionage operations is by the nature of the subject hard to come by, and equally hard to trust. But three valuable sources describe the secret police of Austria: A. Fournier, *Die Geheimpolizei auf dem Wiener Kongress* (Vienna, 1913), M.H. Weil, *Les dessous du Congrès de Vienne* (Paris, 1917), which reprints the letters and gossip which François Hager, head of the police, presented almost daily to his emperor; and Donald Emerson, *Metternich and the Political Police* (The Hague, 1968) shows the ubiquity of clandestine surveillance. See also E. d'Hauterive, *La Police Secrète du Premier Empire* (Paris, 1963–4), Peter de Polnay, *Napoleon's Police* (London, 1970), and on espionage Pellet and Godlewski below.

Correspondence

In addition to useful background material such as that contained in Volumes IX, X, and XI of the *Supplementary Despatches* of the Duke of Wellington (London, 1862), the most important source is *La correspondance de Napoléon Ier* (Paris, 1858–70), Volumes 27, 28, and 31. Most of these letters, and the retrospect on Elba which dates from the St Helena exile, were dictated by Napoleon, and the style reflects the hurried notation of the scribe to hand rather than Napoleon himself. His exchange of letters with Marie-Louise can be found in C.F. Palmstierna, *My Dearest Louise: Marie-Louise and Napoleon 1812–1814* (London, 1958). Philip Guedalla, *The Letters of Napoleon and Marie-Louise* (London, 1935) is not complete, and is in other ways less satisfactory. *The Correspondence of Lord Castlereagh* (London, 1851) and his *Despatches* (London, 1848–53) both contain much illuminating material, as do Prince Murat and Paul le Brethon, *Lettres et documents pour servir a l'histoire de Joachim Murat* (Paris, 1914), *Correspondance Inédite du Prince Talleyrand et du Louis XVIII, pendant le Congrès de Vienne*, edited by L.G. Pelissier (Paris, 1881), and *Talleyrand Intime, d'après sa correspondance avec la Duchesse de Courland. La Restauration en 1814* (Paris, n.d.). The account of Paris in April 1815 sent by John Cam Hobhouse (Lord Broughton) can be found in his *Recollections of a Long Life* (London, 1909–11).

Elba

Descriptive   The island is described in Thiébaut de Berneaud, *Voyage à l'île d'Elbe* (Paris, 1808), which Napoleon consulted at Fontainebleau, in Richard Colt Hoare, *Tour Through the Island of Elba* (London, 1814), H. Barker, *Description of Elba* (London, 1816), A. Pons de l'Hérault, *L'île d'Elbe au début de xixième siécle* (Paris, 1877), and in a modern context by Vernon Bartlett, *A Book About Elba* (London, 1965).

Historical   The books describing Napoleon's stay on Elba vary greatly in reliability and quality, and often rely on undated and dubiously authenticated anecdotes. Count Truchsess-Waldbourg left his account of the journey into exile in his *Relation de l'itinéraire de Napoléon de France a l'île d'Elbe* (Paris, 1815). Other contemporary accounts include the reports of such English visitors as Lord Ebrington, 'Memories of Two Conversations with the Emperor Napoleon' (London, 1839), G.W. Vernon, 'Sketch of a Conversation at Elba' in the *Miscellanies* of the Philobiblon Society, Volume VIII (London, 1863–4), and J.B. Scott, *An Englishman at Home and Abroad* (London, 1930). L.G. Pelissier reprinted contemporary documents in *Le registre de l'île d'Elbe* (Paris, 1897). J. Chautard, *L'île d'Elbe et les Cent Jours* (Paris, 1851) is a useful book, as is the generally reliable H.W. Woolf, *The Island Empire* (London, 1855). Amedée Pichot, *Napoléon à l'île d'Elbe* (Paris, 1873) is another example of the books written to sustain the Napoleonic legend.

E. Foresi, *Napoleone all'isola d'Elba* (Florence, 1884) was an early book using Italian sources. M. Pellet, *Napoléon a l'île d'Elbe* was a landmark in Elban studies, for it was Pellet who found and partly reprinted the Oil Merchant's reports in the French consulate at Livorno, and his account throughout is generally reliable though uneven in texture. P. Gruyer, *King of Elba* (London, 1906) is a good conventional text. Norwood Young, *Napoleon in Exile: I—Elba* (London, 1914) was based on serious research and, in addition to an interesting appendix on contemporary caricature by A.M. Broadley, it contains information not easily found elsewhere. Though outdated in some respects, it remains one of the best works on the subject. A.P. Herbert, *Why Waterloo?* (London, 1952) is one of the oddest and most interesting of all the books on Napoleon's stay on Elba. Though written as a light piece of fictionalized history it was so carefully researched, and contains so many intelligent guesses, that one wishes that Sir Alan Herbert had written a more formal work—and one that was a little less partial to the exiled emperor. P. Bartel, *Napoléon a l'île d'Elbe* (Paris, 1947) gives a straightforward though wordy account of events on the island. Robert Christophe, *Napoleon on Elba* (London, 1964) is a crisp and popular book. It was Christophe who first made the probable identification of Alessandro Forli as the Oil Merchant. Guy Godlewski, *Trois Cents Jours d'Exile* (Paris, 1961) presents much fresh material, especially on clandestine intrigues in and around Elba.

Pons and Campbell These are the two chief witnesses on Elba who describe their impressions in detail. A. Pons de l'Hérault, *Souvenirs et anecdotes de l'île d'Elbe* (Paris, 1897), which was edited by L.G. Pelissier, is lengthy and indispensable, though it is vague about dates, it gives far too much space to the long dispute between Pons and Napoleon about the revenues from the iron mines, and its overblown rhetoric becomes tedious. But Pons was on the whole a good and honest reporter, who was one of the inner circle at the Villa Mulini. His book is the French complement to Neil Campbell, *Napoleon at Fontainebleau and Elba* (London, 1869), published nearly forty years after Campbell died, which contains his journal and also a memoir by his nephew. The journal is the one source which is continuous, contemporary, and dated, and it therefore provides a control on other more anecdotal sources. Campbell's facts are generally trust-worthy, though he was prone to report rumours and to spice his journal entries with his own speculations. But the journal runs very close to the reg-ular and confidential despatches he sent to Castlereagh, now available in the Public Record Office, and in some cases the journal is simply a copy of the current despatch.

Documentary

The Public Record Office, like the Archive National in Paris and the ar-chives of various French ministries, contains much material relevant to Napoleon's stay on Elba and his escape. The log of the *Partridge*, in the PRO, is most useful. So are Lord Burghersh's comments on Campbell's failing duty (FO 79/21, 22, 23), letters relating to Campbell's dispute with Captain Tower (FO 27/114, May 1814–April 1815), and general back-ground material listed under Tuscany and Rome (FO 170, 171), Genoa (FO 762, 1815); there are also many reports and letters dealing with the conduct of diplomatic business in France and Vienna. There are interest-ing letters written by Campbell to Lord William Bentinck in the Portland Papers in the University of Nottingham Library (PwJd 1154, 1156, and 1158) and PwJd 1159 is a set of notes kept by Campbell on the *Partridge* during the vain voyage to Antibes. There is a letter from Campbell to Lord Bathurst in *HMC 81 Supplementary Report on the Manuscripts of Robert Graham Esq. of Fintry* (London, 1942). Other manuscript sources include a letter from Captain Ussher to Captain Waldegrave, which is in the library of the National Maritime Museum at Greenwich, as are the letter-boxes of Admiral Pellew. Lord Liverpool's papers may be found in the Manuscript Room of the British Library. Most of Lord Castlereagh's papers are in the Public Office in Belfast and the County Archive Office in Durham. There are some papers and other useful documents in the Villa Mulini and in the public library at Portoferraio. The debates in Parliament after Napoleon's escape are full of interest and evasions. See House of Commons 6, 7, 20, 22, and 25 April 1815, and House of Lords, 12 April 1815, especially the attempts of Castlereagh and Liverpool to justify their ambiguous policy and the orders given to British ships in the vicinity of Elba.

INDEX

JAN 2 8 1903